Science Projects in a Pocket

Macmillan/McGraw-Hill

Instruction For Copying

Answers are printed in non-reproducible blue. Copy pages on a light setting in order to make multiple copies for classroom use.

Contents

Contents

LIFE SCIENCE: PLANTS AND ANIMALS

EARTH SCIENCE: SEASONS

Contents

EARTH SCIENCE: RESOURCES AND FOSSILS

EARTH SCIENCE: PLANETS, MOONS, AND STARS

Contents

PHYSICAL SCIENCE: MATTER AND MOVEMENT

PHYSICAL SCIENCE: ENERGY

How to Use Science Projects in a Pocket

Science Projects in a Pocket is intended to engage children in learning the key concepts of life science, earth science, and physical science. Through these exciting, creative, and easy-to-follow activities, first- and second-grade children can build upon the basic foundations of science and create portfolios for their future reference and study.

This book is comprised of activities that focus on developing children's reading, writing, music, art, and math abilities while strengthening their science skills. Each pocket of activities in this book contains four projects to enhance children's understanding of a specific science topic. Children should begin by making a pocket folder for each set of pocket activities in this book. The pocket folder contains four pockets into which they can place their completed projects.

With teacher guidance, children will create projects that utilize their creativity as well as their science knowledge. Activity sheets, which are referenced in the individual activities, accompany many of the activities and can be found in the second half of this book. These activity sheets should be reproduced and utilized by children to complete the various pocket activities.

This pocket-folder format will allow children to easily review the concepts studied in class and enable them to share their various science projects at home. The activities in this book are simple to prepare, present, and complete. It is a natural extension of the A to Z Activity Book designed for use in the kindergarten classroom, and it provides many opportunities for enhanced science instruction and comprehension in the first- and second-grade classrooms. Enjoy!

How to Create a Pocket Folder

Children should create one pocket folder for each set of four pockets presented in this book. Pocket folders are created by using three sheets of 12" x 18" pieces of construction paper and a stapler.

Step 1: Have children fold one sheet in half vertically.

Step 2: Children should slide another sheet in between the folded halves of the sheet folded vertically, and then fold the booklet in half horizontally. This step forms four pockets in which to put completed activities.

Step 3: Have children staple the outer sides where the two pieces of paper are together. This will seal the outer sections of the pockets.

Step 4: Next, children should fold the third sheet horizontally and place it around the booklet to create a cover. Then, they should staple the entire booklet along the spine to hold the pages together.

Step 5: Finally, children can decorate each pocket folder by using construction paper, scissors, glue, and the pocket labels that are located in the resource section of this book.

How to Use the Activities in This Book

The activities in this book have been designed as learning aids for children. Some children may have difficulty completing all of the tasks required, while others may want to go beyond the requirements of the activities. Safety is also important to keep in mind since children will often be required to use scissors. Below are some suggestions for implementing the activities presented in this book.

Safety

- Scissors are used in many of the projects. While many children understand how to use scissors, a discussion on safety should precede the use of any activities that require the use of scissors.
- Paints and markers should be non-toxic and washable.
- Clean-up is an important part of any activity. Safety should always be kept in mind when instructing children to assist you in cleaning.

Creating Shapes and Patterns

- Have several objects available for tracing to create such shapes as circles, squares, and rectangles.
- Assist children in cutting when necessary, or provide pre-cut shapes as an option for children who are having difficulty.

Background Research

Several activities suggest that you lead a discussion based on the material discussed within the activity. You may wish to consult research books and online resources as helpful ways to obtain more information on the topics.

Modifying Activities

These projects were designed for first- and second-grade skill levels. The projects can be modified based on the varying abilities of the children in your class. You can modify the activities so that they are more challenging for those whose skills warrant a higher level of difficulty. Consider having children conduct research or write several sentences about their projects. Other options include making additions to the activity, or teaching smaller groups of children about what they have learned about the various topics.

Life Science

Language Arts — Flower Parts

Materials
white paper, markers, scissors, small pieces of construction paper in the following colors: purple, green, yellow, manila

Objective: Review parts of plants.

Science Inquiry Skills: identify, communicate, collect data

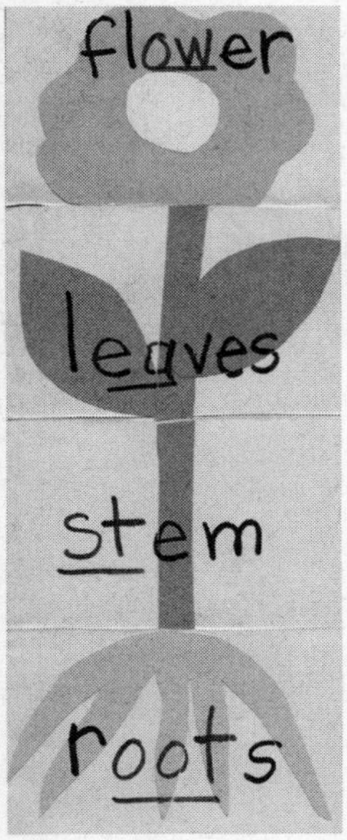

- Make a list on the board of the parts of a flower: *root, stem, leaves, flower*. Provide white paper to children and direct them to fold the paper lengthwise. Next, have children fold the paper in half from top to bottom, and from top to bottom again.
- Have children open up the paper and cut along the three folds on the left-hand side. Tell children that they should only cut to the middle of the paper so that four flaps are created. Have children re-fold the paper lengthwise and draw roots on the bottom flap. Continue with the stem, leaves, and flower in each section.
- Label the sections with the following clue words: *r**oo**ts, **st**em, l**ea**ves, and fl**ow**er*. Have children open each flap and write more words that have the same sounds as the underlined part of the clue word. For example, under the flower picture, children can write *how, town, now, plow, clown, frown,* and *brown*.
- Children in first grade could write words that begin with the same sound as the clue words.

Science — Eating Different Parts of Plants

Materials
markers or crayons

Objective: Identify the root, stem, leaves, and flower parts that are edible.

Science Inquiry Skills: identify, sort, classify, communicate

Resources: "Eating Different Parts of Plants" activity sheet, p. 69

- On the board, develop a list of fruits and vegetables that we eat. Help children decide if the edible part is the root, stem, leaves, or flower. Write the edible part next to each fruit and vegetable on the list.
- Provide a copy of the activity sheet to each child. Have children interview some classmates, asking them to name their favorite fruits and vegetables. Children should record these answers on the activity sheet under the appropriate plant part. Finally, instruct children to draw the fruits and vegetables.

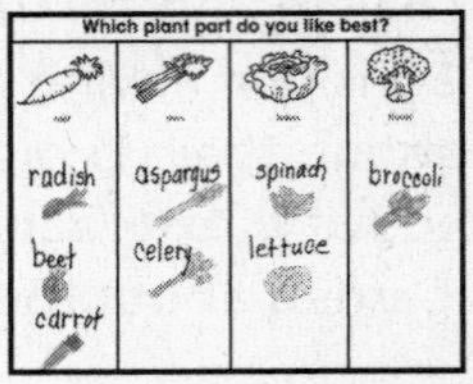

Music

What Is the Part?

Materials

construction paper in assorted colors, glue, scissors, markers

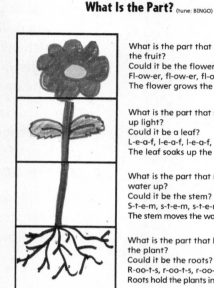

Objective: Identify the parts of plants.

Science Inquiry Skills: identify, communicate, make a model

Resources: "What Is the Part?" activity sheet, p. 70

- Write the words to the song "What Is the Part?" on the board. Teach the song to the tune "Bingo."
- Provide each child with a copy of the activity sheet, which includes the song and four sections. Have children create four parts of a plant in the sections that correspond to the song. Start with the roots in the bottom section, followed by the stem, leaves, and flower.

What Is the Part? (tune: BINGO)

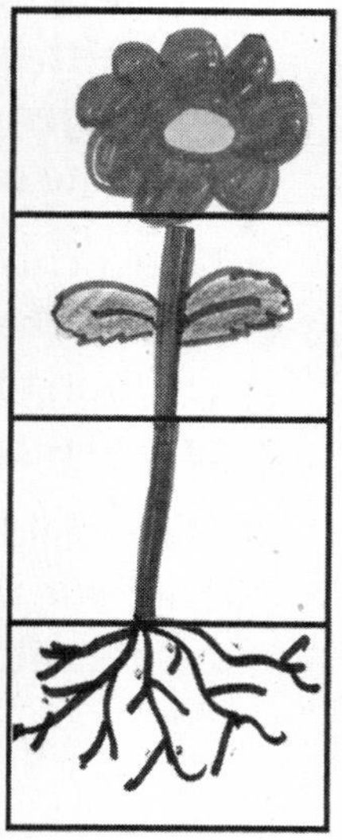

What is the part that grows the fruit?
Could it be the flower?
Fl-ow-er, fl-ow-er, fl-ow-er,
The flower grows the fruit.

What is the part that soaks up light?
Could it be a leaf?
L-e-a-f, l-e-a-f, l-e-a-f,
The leaf soaks up the light.

What is the part that moves water up?
Could it be the stem?
S-t-e-m, s-t-e-m, s-t-e-m,
The stem moves the water up.

What is the part that holds the plant?
Could it be the roots?
R-oo-t-s, r-oo-t-s, r-oo-t-s,
Roots hold the plants in place.

Art

Flower Parts Quilt

Materials

8" x 8" pieces of light-blue and green construction paper, flower wrapping paper, cupcake paper, green pipe cleaners, brightly colored pom poms, green tissue paper, tan raffia pieces, brown and yellow construction paper, black markers

Objective: Display parts of a plant.

Science Inquiry Skills: make a model, recognize patterns

Resources: "Flower Parts Quilt" activity sheet, p. 71

- Review the four parts of plants with children and write the words on the board. Direct children to tear brown construction paper to create soil in the bottom section of the square.
- Next, have children create a flower by using raffia for roots, a green pipe cleaner for a stem, green tissue for leaves, and cupcake paper and pom poms for a flower. Ask children to use a black marker to label each section and add black "stitching" around the sides.
- Cut ten 2-inch squares of flower wrapping paper and two 2-inch pieces of yellow construction paper for each child. Provide the quilt pattern to children and direct them to fill in the marked areas with wrapping paper with yellow paper in the middle. Some of the squares will need to be cut into triangles.

Science — Flower Wheel

Materials
orange, white, and yellow construction paper, markers, brads, scissors

Objective: Identify what plants need to live.

Science Inquiry Skills: identify, communicate, make a model

- Discuss the needs of plants and highlight the words *soil, water, air,* and *light*. Explain to children that they will create a model of plants' needs.
- Have children cut out two equal circles—one from yellow construction paper and one from white paper. Ask children to divide the white circle into four equal parts and write the words *soil, water, air,* and *light* in the sections. Then have children illustrate one word in each of the sections.
- Direct children to cut out a $\frac{1}{4}$ piece from the yellow circle and write "What do plants need?" on the remaining $\frac{3}{4}$ section. Ask children to stack the yellow circle on top of the white circle and attach the two with a brad through the center of the circles.
- Have children cut petal shapes out of red construction paper and glue them around the perimeter of the yellow circle to create a flower.

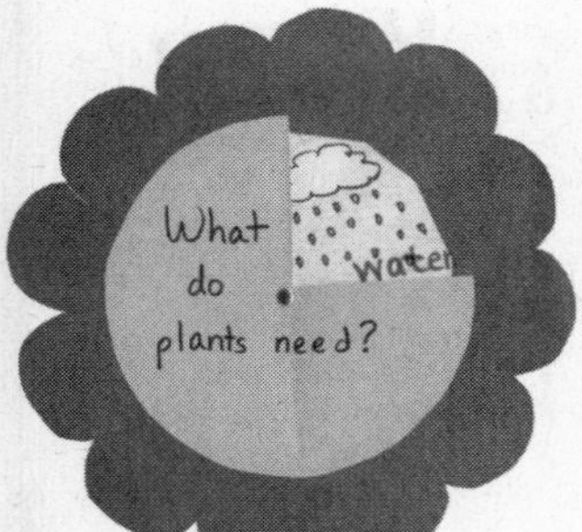

Science — Flower Glyph

Materials
glue, black markers, 8" x 8" pieces of light-green construction paper, pieces of paper in the following colors: dark blue, light blue, dark brown, light brown, orange, yellow

Objective: Make a model of the needs of plants.

Science Inquiry Skills: identify, communicate, sort, classify

Resource: "Flower Glyph" activity sheet, p. 72

- Provide children with 8" x 8" squares of light-green construction paper and have the following colors of construction paper available: dark blue, light blue, dark brown, light brown, orange, and yellow.
- Read the legend questions to children to help them build a flower picture. Children should illustrate a picture that shows soil, water, light, and air, based on their responses to the questions. Have children add black "stitching" around the side of the square.
- After children complete their pictures, help children sort their pictures into two columns on the floor as a class, based on the first question on the legend. Continue with each question in the legend. Save this glyph to be used with the "Flower Quilt" activity.

Art

Flower Quilt

Materials

8″ x 8″ pieces of white construction paper, 4″ x 4″ pieces of tan, yellow, orange, and dark-blue construction paper, scraps of orange, yellow, brown, and red paper, white paint, black marker

Objective: Create a quilt displaying the needs of plants.

Science Inquiry Skills: compare, recognize patterns

- Give children squares of tan, yellow, orange and dark-blue construction paper. Have children add glue to the bottom of a tan square and sprinkle coffee grains on it to make soil. Then ask children to draw a watering can on the yellow square with drops of water coming out, and a yellow Sun on the orange square. Have children dip a small lid in white paint to create a stamp to form air bubbles on the dark-blue square.
- Direct children to glue the 4″ x 4″ pieces onto white construction paper, placing the tan and yellow squares above the orange and dark-blue squares. Have children label the appropriate squares with the words *soil*, *water*, *air*, and *light*. Then have children draw stitches around the edges of the squares.
- Help children make a quilt. To form the quilt, have children glue each square that they constructed to a large piece of butcher paper. The squares should be glued from left to right. Help children find patterns in the quilt.

Science

The Needs of Plants

Materials

sentence strips or 4″ x 18″ construction paper, markers, 3″ squares of white paper

Objective: Identify the needs of plants.

Science Inquiry Skills: sort, classify, identify, communicate

- Review the needs of plants with children. Use suggestions from children to list the needs of plants on the board.
- Provide children with 3″ squares of white paper. Have children draw pictures of needs and non-needs of plants. Children should then sort the needs into one pile and non-needs into another pile.
- Have children fold a sentence strip in half and then unfold the sentence strip. Children should write "Plants need" on the left side and "Plants do not need" on the right side. Finally, have children staple the appropriate pictures to the right of the sentences.

Science

Plants Grow Everywhere

Materials
markers or crayons, scissors, stapler

Objective: Identify plant environments.

Science Inquiry Skills: compare, identify, classify

Resources: "Plants Grow Everywhere" activity sheets, pp. 73–74

- Duplicate the activity sheets back-to-back for children, and direct them to fold the sheet in half vertically and horizontally, forming a four-page booklet. The title should form the front cover, and the description of the cattail should form the back cover. Students should then cut along the center horizontal fold, and staple the pages together along the vertical fold.
- Have children read the descriptions of plant environments in the booklet.
- Finally, have children draw illustrations above *hot*, *cold*, and *wet*. For example, children can draw the Sun, snow, or rain.

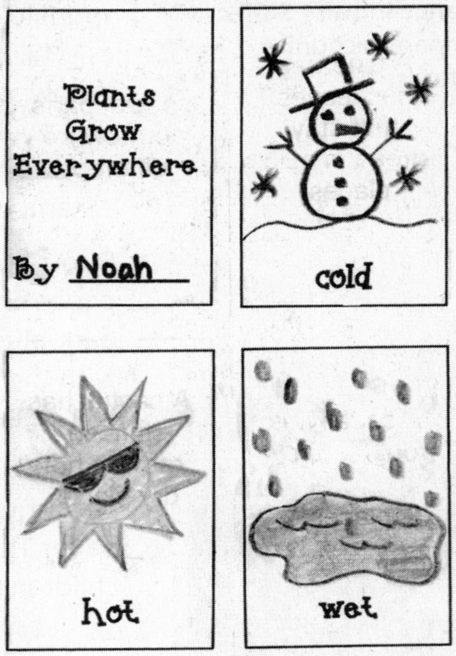

Art

Plant Environments

Materials
9″ x 9″ blue construction paper, glue, crayons or markers

Objective: Make a display of plant environments.

Science Inquiry Skills: compare, identify, communicate

- Have children fold a square piece of paper diagonally to form a triangle and then fold it again to form a smaller triangle. Next, have children unfold the paper and cut along one of the folds to the center of the triangle. Place a small amount of glue along the fold opposite the cut and re-fold the paper into a triangle.
- Have children draw three environments on the sides: a desert scene, a lake with plants, and pine trees in snowy mountains. The triangle shape can then be propped up into a pyramid.

Science

Living Things Grow Everywhere

Objective: Identify plants and their environments.

Science Inquiry Skills: sort, classify, identify

- Have a discussion with the class about the environments in which plants grow. Write *cold*, *wet*, and *hot* on the board and guide children in listing the plants that grow in each environment.
- Provide children with sentence strips and 3" white construction paper squares (6 or more per child). First, have children write *hot*, *cold*, and *wet* on three squares. On the three remaining squares, direct children to draw and label three plants, one for each environment.
- Have children write, "A ______ grows where it is______." on a sentence strip. Direct children to staple their environment squares and plant squares to the appropriate spaces on the sentence strips.

Materials

sentence strips or 4" x 18" construction paper, white construction paper or copying paper

Language Arts

Plant Cube Toss

Materials

wooden cubes, black markers

Objective: Spell words that begin with the letters in the word "plants."

Science Inquiry Skills: organize data, communicate, identify

Resources: "Plant Cube Toss" activity sheet, p. 75

- Provide each child with the activity sheet, a wooden cube, and a black marker. Instruct children to write one letter from the word "plants" on each side of their cubes.
- Have each child roll a cube, check the letter displayed, and write a word beginning with that letter in the square below that letter on the activity sheet. Direct children to continue rolling the cubes and writing words until one or more columns are filled.

p	l	a	n	t	s
peanut	lime				seed
pod	lemon			tree	sprout
pumpkin	limb			tomato	soil
peas	leaf	apple	nuts	turnip	stem

Science — Plants Change

Materials
orange and green construction paper, crayons or markers

Objective: Identify a sequence of plant growth.

Science Inquiry Skills: observe, identify, make a model

- Discuss with children how plants grow and change. List several plants that begin as seeds and review their life cycles.
- Give children green construction paper. Have them trace one of their hands on the paper and cut it out.
- Next, give each child a piece of orange construction paper and instruct them to fold it in thirds, like a letter. While the paper is folded, have children cut out a carrot shape like the one on the left. They should then glue the green cut-out to the top of the carrot.
- Children should then unfold their carrots and draw a three-stage carrot life cycle inside it. The first section should have carrot seeds in soil. The middle section should have water and sunshine being added, and the last section should have a child eating a fully grown carrot. Finally, have children write a title on the front of the booklet.

Science — Apple Tree Logic Line-Up

Materials
crayons or markers

Objective: Identify a plant growth sequence.

Science Inquiry Skills: identify, communicate, interpret data, draw conclusions

Resources: "Apple Tree Logic Line-Up" activity sheets, pp. 76–77

- Provide each child with a copy of the activity sheets and direct them to color and cut out the four cards. Enlarge, color, and laminate one set of cards for use as a class display. The four picture cards show the stages of fruit growth on an apple tree. The illustrated stages are a tree with leaves, a tree with flowers, a tree with apples, and an apple displaying the seeds.
- Ask four children to stand in front of the group; each one should hold one of the cards. For each problem on the first activity sheet, read the steps aloud and have the class arrange the four children holding the large cards to the match problem. Children should rearrange their cards at their desks to follow the class.

Art

Blueberry Growth Cycle

Materials
light-blue construction paper, blue paint, paintbrush, crayons or markers

Objective: Identify the sequence of plant growth.

Science Inquiry Skills: observe, identify, organize data

- Discuss the uses of various plants and the environments in which they grow. Have children take turns naming a fruit or vegetable, describing where it is grown, and listing the products made from the plant.
- Give children light-blue construction paper and blue paint. Have children fold the paper in half lengthwise, then fold it in thirds the other way. Direct children to open the paper and cut along two folds on the top section, making three flaps.
- Instruct children to create illustrations on the outer flaps of the paper. They should draw and label a blueberry bush on the left section, a basket of blueberries in the middle section, and a product made with blueberries, such as a muffin, in the right section.
- Finally, have children lift each flap and write one or more sentences describing their pictures.

Language Arts

Plants Spin and Spell

Materials
scissors, crayons, pencils, brads, paper clips

Objective: Identify the parts of plants that grow and change.

Science Inquiry Skills: identify, communicate, organize data

Resources: "Plants Spin and Spell" activity sheets, pp. 78–79

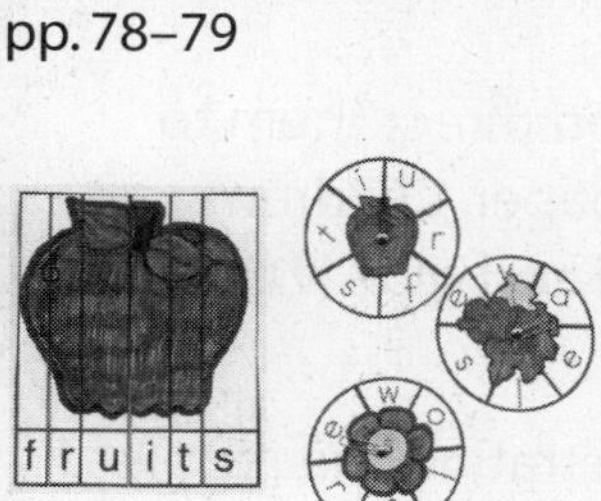

- Duplicate the first activity sheet for each child. Have children cut the pictures into six strips each.
- Review the parts of the plants that grow and change—*leaves*, *flowers*, and *fruits*.
- Make three plant spinners by placing a paper clip and brad in the middle of each plant spinner circle.
- Stand in front of the class and spin a plant spinner. Write the selected letter on the board. Have children list as many plant words as they can that begin with the selected letter. Continue spinning until every letter has been selected. Children can arrange their strips to keep track of which letters have been used.

Art

Birds Are Animals

Materials

yellow, orange, blue, black, white, and pink construction paper, paper plates, yellow feathers, white 7" circles of paper, markers, scissors, glue, rulers

Objective: Identify birds and their environments.

Science Inquiry Skills: sort, classify, identify, communicate

- With class participation, develop a list of birds that fly, walk, and swim. Discuss how they get food and shelter.
- Have children fold and glue paper plates in half. Next instruct children to cut a circle from yellow paper for the head of the duck, a bill shape from orange paper and webbed feet from black paper. They should glue these onto the plate, making a duck. Direct children to draw eyes and glue yellow feathers to the body and head.
- Give each child three white paper circles. They should fold them in half and cut slots $1\frac{1}{2}$" apart and $1\frac{1}{2}$" deep, forming three pop-out sections. Children should unfold the papers and write "Some birds fly" on the first section, "Some birds swim" on the second section, and "Some birds run" on the third section.
- Children should make three birds out of construction paper. They should glue one bird to each pop-out section and draw either clouds, water, or land on the background. The circles should be glued inside the duck's body.

Science

Reptiles Are Animals

Materials

markers, various colors of construction paper, glue, light-blue paper, scissors

Objective: Identify reptiles in their environments.

Science Inquiry Skills: organize data, identify, communicate

Resources: "Reptiles Are Animals" activity sheet, p. 81

- Hold a class discussion about the various types of reptiles with a focus on how their needs for water, food, and shelter are met. Each child should pick a reptile as a focus for this pocket.
- Give each child three pieces of blue paper. They should stack the pieces, move the bottom sheet up 1", and then the middle sheet up 1". They should fold the papers in half so that a booklet with six graduated levels is created.
- Give each child a copy of the activity sheet and direct them to cut out the sentence trips and glue them to paper. Children should draw their reptile in the oval provided and glue it to the paper.
- Children should complete each sentence, illustrating the reptile in each situation.

Science

Mammals Are Animals

Materials

white or light-blue paper, markers

Objective: Identify mammals and their environments.

Science Inquiry Skills: identify, sort, classify, communicate

- Conduct a class discussion about how mammals meet their needs, and develop a list of mammals and their environments.
- Provide paper and instruct children to fold the paper in half by width and then by length. Once opened, there will be four sections.
- Direct children to label the sections as *grasslands, desert, jungle, forest*, and have children illustrate each section with an animal that lives in each environment.
- Children should write a statement in the correct section about each animal.

Language Arts

Ocean Creatures Are Animals

Materials

markers, crayons, stapler

Objective: Identify ocean animals.

Science Inquiry Skills: identify, communicate, classify

Resources: "Ocean Creatures Are Animals" activity sheets, pp. 82–83

- Provide time for the children to compile a list of animals that live in the ocean.
- Duplicate the activity sheets for children, making one copy of the cover and five copies of the inside sheet for each child. Give children the pages and have them form a booklet from the sheets. Discuss the ocean animals on the back page.
- Next, children can illustrate the ocean creatures from the back cover and write sentences about each animal. When children are finished, have them staple their booklets.

Art

Ocean Environment

Materials
white paper, watercolor paints, brushes, crayons, black pens

Objective: Identify the ocean as an environment for some animals.

Science Inquiry Skills: identify, organize data, communicate

- Develop a list with children of animals that live in the ocean. Show them pictures of the ocean animals.
- Direct children to draw and color pictures of ocean animals on white paper. They can add sand, water, and plants.
- After completing the drawings, children should paint a watercolor wash over the pictures. When the drawings are dry, children can write the names of the animals in black pen on the paintings.

Art

Desert Environment

Materials
white paper, watercolor paints, brushes, construction paper in assorted colors, tissue paper in assorted colors, markers, scissors, glue

Objective: Identify the desert as an environment for some animals.

Science Inquiry Skills: identify, observe, communicate

- Brainstorm with children the names of desert animals and write them on the board. They can use picture books to help them.
- Give children white paper and instruct them to paint a watercolor background with layers of various colors that represent a sunset or sunrise.
- Have children create mountains, cacti, and a Sun from construction paper and glue it to the painting. They should use the tissue paper to make cactus flowers, and markers to add detail.

Art

Mountain Environment

Materials
blue construction paper, paints, brushes

Objective: Identify mountains as an environment for some animals.

Science Inquiry Skills: identify, communicate, display data

- Discuss the mountain environment with children and help them develop a list of animals that live in the trees and mountains.
- Provide children with blue construction paper and demonstrate painting trees on it with long, wide strokes.
- Children should paint three or more trees in their picture and then draw animals near the trees.

Art

Jungle Environment

Materials
wide green ribbon, construction paper in various colors, stapler, markers, scissors

Objective: Identify jungles as an environment for some animals.

Science Inquiry Skills: sort, classify, identify

- Discuss the jungle environment with children and help them develop a list of small animals and plants that live there.
- Help children use markers to draw leaves, flowers, snakes, and insects on construction paper. Then, provide children with green ribbon. Have each child cut out their drawings and staple them along the length of the ribbon to create a jungle vine.

Science

Water Animals Are Different

Materials
9″ x 12″ gray construction paper, markers, scissors

Objective: Compare two water animals.

Science Inquiry Skills: sort, identify, compare

- Discuss with children the similarities and differences between whales and fish. Some aspects to compare are their habitiats, diets, young, and how they breathe. List the differences on the board.
- Provide children with 9″ x 12″ gray construction paper and direct them to fold it in half lengthwise. Have them cut a line down the middle of one fold to make two flaps. Children should draw a whale on one flap and a fish on the other.
- Have children lift the flaps and write several sentences about whales and fish underneath the appropriate flaps. For example, children may write, "Fish breathe through gills," and "Whales breathe with lungs."

Science

Animal Shelters

Materials
blue construction paper, markers, scissors

Objective: Identify four types of shelters for animals.

Science Inquiry Skills: sort, classify, identify, communicate

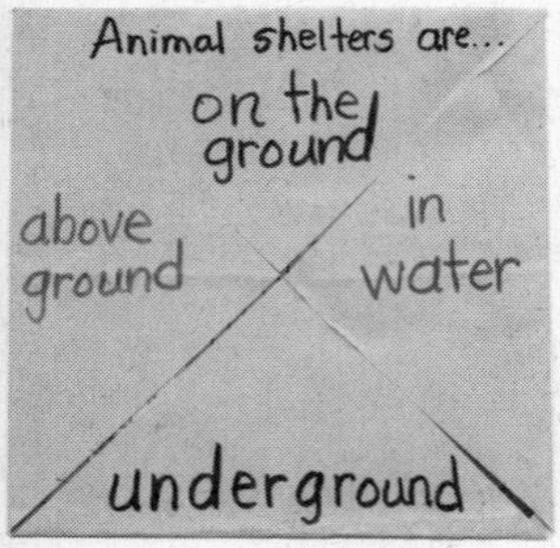

- Ask children to identify various shelters for animals and list the locations of these shelters. On the board, write the labels *on the ground*, *underground*, *above ground*, and *in water*. Next, ask children to identify animals and shelters in those locations.
- Provide children with 9″ x 9″ blue construction paper and instruct them to fold each corner inward until the four corners meet in the middle.
- On the front of each triangle fold, direct children to write one of the following: *on the ground*, *underground*, *above ground*, or *in water*. "Animal shelters are..." should be written on the top flap to accompany one of the phrases. On the inner portion of each fold, instruct children to write a sentence about each location, such as, "Grass and rocks are shelters for animals on the ground."
- Finally, instruct children to open each fold and illustrate each location in the blank square portion of the paper.

Music

Animals and Their Foods

Materials
$8\frac{1}{2}$" x 11" brown and white copy paper, plastic baggies, markers, scissors, stapler

Objective: Identify foods that animals eat. Focus on food needs of bears.

Science Inquiry Skills: identify, communicate, organize data

Resources: "Animals and Their Foods" activity sheets, pp. 84–85

- Duplicate the bear pattern on brown paper for each child. Give each one a copy of both and activity sheets. Teach children the song, similar to "I Know an Old Lady."
- Children should color the food items on the activity sheet, write names for the food items, and cut them out. They should also cut out the bear and the circle on the bear from the sheet. Assist them in stapling a baggie to the back of their bear.
- Children should sing the song and put the food in the baggie as sequenced in the song. Children can brainstorm about other animals and what those animals might eat.

Science

Lions and Lambs

Materials
black, brown, and yellow construction paper, white paper, markers, scissors, cotton, glue

Objective: Identify and compare needs of two animals.

Science Inquiry Skills: sort, classify, identify, compare

- Have a discussion with children about the needs and shelters of animals, and make a list of food needs and shelters.
- Provide children with 9" x 12" white construction paper and direct them to fold the paper in half lengthwise and then into three equal sections. Ask children to open the paper and cut along the folds on the top half of the paper.
- Have children create a lamb head from the first flap section by cutting its free edges in a wavy line. Children should glue cotton "wool" on top of the lamb head, create ears out of black paper, and draw eyes, a nose, and a mouth. In the second section, have children write "Both" on the flap. On the third flap, have children create a lion's head by rounding the corners of the flap, adding a mane with strips of brown paper, and using yellow paper to create ears, adding features with a marker.
- Finally, have children write statements about the animals under the appropriate flaps. In the middle section, have children write statements relevant to both animals.

Science

Animal Movements

Materials
drawing paper, markers, pencils, stapler

Objective: Identify methods of movements for various animals.

Science Inquiry Skills: identify, sort, classify, make a model

Resources: "Animal Movements" activity sheet, p. 86

- Lead a discussion with children about movements of animals. Put the labels *Land*, *Air*, and *Water* on the board and ask children to name animals and decide whether the animals move on land, in the air, or in water. Children may use library books to list more animals.
- Provide children with drawing paper and the "Animal Movements" activity sheet. Have children color and cut out the "Animals Move" top and staple it to their paper. Under each heading, children should draw and label animals for each environment.

Animals Move
on land
in air
in water
chicken
rabbit
goat
owl
eagle
parrot
fish
dolphin
jellyfish

Science

Animal Teeth

Materials
gray, green, white, and pink construction paper, small marshmallows, scissors, glue, crayons

Objective: Identify the types of teeth that plant- and meat-eating animals have.

Science Inquiry Skills: identify, observe, compare

Resources: "Animal Teeth" activity sheets, pp. 87–88

- Duplicate the "Animal Teeth" activity sheets, copying the hippo body on gray construction paper, the alligator body on green construction paper, and the mouths on pink paper.
- Show children pictures of different types of animals and discuss the foods the animals eat. Help children realize that meat-eating animals have sharp teeth to tear meat, and plant-eating animals have flat teeth to grind plants.
- Provide children with the hippo patterns and direct them to cut out the pieces, then fold and glue the mouth onto the face of the hippo. For teeth, children can glue marshmallows on the top and bottom of the mouth. Inside the mouth, children should write sentences about plant eaters.
- Next, give children the alligator patterns to cut out and glue together. Children should draw jagged, pointed teeth on white paper and then cut out and glue them inside the mouth. Children should write statements about meat-eating animals inside the mouth.

Science

Animal Coverings

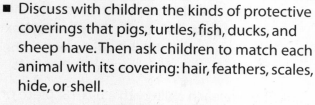

Materials
construction paper in various colors, manila paper, scissors, glue, markers, stapler

Objective: Identify types of animal coverings.

Science Inquiry Skills: sort, classify, identify, make a model

Resources: "Animal Coverings" activity sheet, p. 89

- Discuss with children the kinds of protective coverings that pigs, turtles, fish, ducks, and sheep have. Then ask children to match each animal with its covering: hair, feathers, scales, hide, or shell.
- Children should select colored paper to match each animal. Give each child a copy of the "Animal Coverings" activity sheet. Children should cut out the oval and use it as a template by tracing its shape on each piece of colored paper they chose. They should then cut out each oval.
- Have children turn the colored ovals into animals by adding identifying elements, such as a head, feet, and a tail. Children should label their manila oval "Animal Coverings." Ask children to write a sentence about the animal and its protective covering on each animal.
- Help children staple the set of animals together, with the manila oval on top as a cover.

Science

Animal Life Cycle

Materials
9" x 9" green paper, light-blue paper, coffee filters, tissue paper in assorted colors, 2" x 3" black paper, hole reinforcements, glue, scissors, markers

Objective: Identify the life cycle of a frog.

Science Inquiry Skills: make a model, identify

Resources: "Animal Life Cycle" activity sheets, pp. 90-91

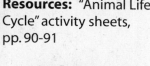

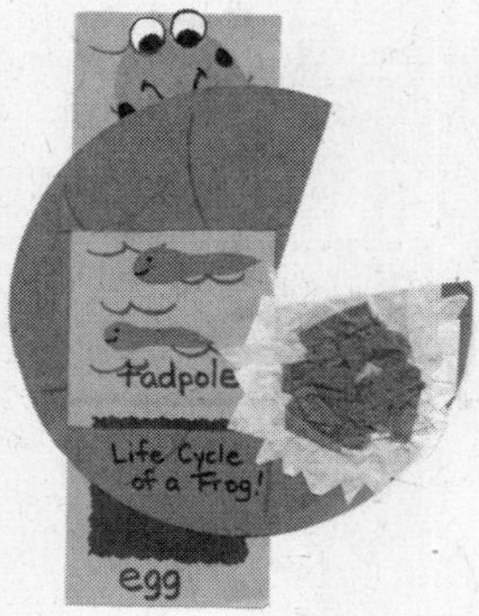

- Give each child a copy of the first activity sheet. Have children cut out the lily pad, trace it on green paper, and cut it out. Next, children should make slits in the green lily pad by cutting along the dotted lines. Children should create a flower using a coffee filter and tissue paper, glue it to the lily pad, and write "Life Cycle of a Frog" next to it.
- Give each child the second activity sheet, copied on light-blue paper, and direct them to cut out the rectangle and fold it into three equal sections. Children should add reinforcements to a black square of paper and glue it to the bottom section, draw two tadpoles in water on the middle section, and draw a frog and water on the top section. Next, have children label the drawings. Children should carefully slip the strip through the slits in the lily pad.

Science

Grasslands Habitat

Materials
6″ x 18″ light-blue paper, markers, scissors, glue, popsicle sticks

Objective: Identify animals that live in the grasslands.

Science Inquiry Skills: identify, communicate

Resources: "Grasslands Habitat" activity sheet, p. 93

- Give each child a copy of the activity sheet and instruct them to cut out the phrases. Next, they should color all of the animals on the sheet and cut them out.
- Provide each child with a 6″ x 18″ piece of light-blue paper. They should fold it in half and in half again to create four equal sections. They should unfold it, glue one phrase in each section, and illustrate the phrase.
- Next, children should glue an animal to each section. They can then glue one animal to a popsicle stick so that it can "travel" through the landscapes.

Art

Forest Habitat

Materials
green paper, 3″ x 9″ brown paper, markers, scissors, glue

Objective: Identify animals that live in forest trees.

Science Inquiry Skills: identify, communicate

Resources: "Forest Habitat" activity sheet, p. 94

- Guide a discussion about animals that live in the forest, focusing on those that make their homes in trees.
- Have each child trace his or her hand on three pieces of green paper and then cut out the hand shapes. They should use the paper scraps to make a leaf bed. Instruct children to create a tree by gluing the hand shapes to the top of a 3″ x 9″ piece of brown paper and adding the leaf bed at the bottom.
- Give each child a copy of the activity sheet. Direct them to color and cut out the animals. They should glue the animals to the section of the tree in which they live: the bottom of the trunk, the trunk, or the leaves.

Art

Water Habitat

Materials

paper plates, blue paint, brushes, blue or clear plastic wrap, construction paper in assorted colors, markers, fun foam, stapler,

Objective: Identify animals that live in water habitats.

Science Inquiry Skills: identify, communicate

- Guide a discussion about ocean animals. Make a list of animals that live in the ocean and in other bodies of water.
- Give each child two paper plates, blue paint, and brushes. Have children paint one plate blue. Then, assist children in cutting out the middle section of the unpainted paper plate.
- Children should make ocean animals from construction paper or fun foam, using markers to draw details. They should glue the animals onto the middle section of the painted plate. Have children label the animals. Then the section should be covered with clear or blue wrap.
- Have children place the unpainted plate's outer section over the painted plate so that it frames the middle section. Assist children in stapling the plates together. Children should label the project "Underwater Habitat."

Science

Rain Forest Habitat

Materials

markers, scissors, stapler

Objective: Identify plants and animals that live in the rain forest.

Science Inquiry Skills: identify, communicate, sort, classify

Resources: "Rain Forest Habitat" activity sheet, p. 95

- Show children pictures of the rain forest. Discuss the different layers of the rain forest and the plants and animals that live in each layer.
- Give each child a copy of the activity sheet. Children should label, color, and cut out the pictures. They should organize the pictures into two piles: plants and animals.
- Have children create 4 sentence strips like the one shown below. Assist children in stapling each pile to the correct space on the sentence strip. The animals should be placed in the first space and the plants in the second.

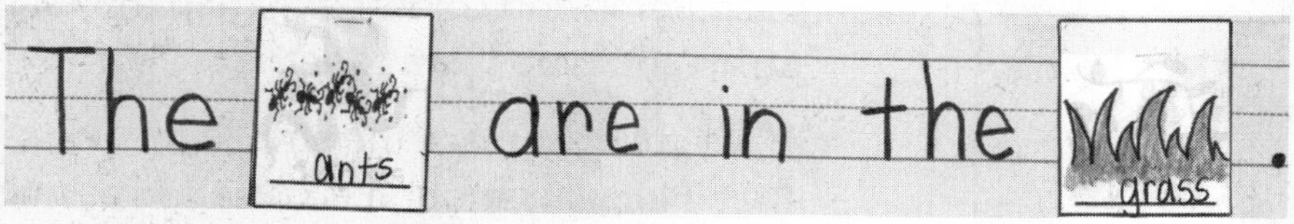

Music

Snakes Grow and Grow

Materials
markers, rulers, stapler

Objective: Learn about snakes of various sizes.

Science Inquiry Skills: identify, communicate

Resources: "Snakes Grow and Grow" activity sheets, pp. 96–99

- Give each child a copy of the activity sheets, copied back to back. Discuss the animals that snakes eat. Teach children the first part of the "Slithering Snake" chant. Then discuss with children the different sizes of snakes described on the first activity sheet.
- Have children make booklets. Children should fold the activity sheets in half and staple them together in the middle.
- In the booklet, children should complete the chant by filling in the descriptive words on each page. They should then illustrate the chant with the correct snake for each page, according to the measurements specified on the back of the booklet. They should then color their pictures.

Language Arts

Meat-Eating Animals

Materials
$8\frac{1}{2}$" x 11" yellow construction paper, markers, scissors, glue

Objective: Identify animals that eat meat.

Science Inquiry Skills: identify, sort, classify

Resources: "Meat-Eating Animals" activity sheets, pp. 100–101

- Lead a discussion about the differences between meat eaters and plant eaters. Ask children to decide whether lions are meat eaters or plant eaters. Give them information about a lion's diet.
- Give each child a copy of the first activity sheet, which should be copied on yellow paper. Instruct them to add details with markers. Children should cut out the lion and make a 3" slit in the lion's mouth.
- Give each child a copy of the second activity sheet. They should decide which food items to draw in each day's box, draw the food items, and color them. Then they should cut out the two strips and glue them together.
- Have children place the strip in the lion's mouth and move it so that it looks as though he is eating each item.

Science

Food Chains Begin With the Sun

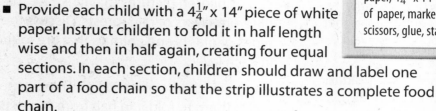

Materials
6" x 6" pieces of yellow paper, $4\frac{1}{4}$" x 14" pieces of paper, markers, scissors, glue, stapler

Objective: Identify a food chain beginning with the Sun.

Science Inquiry Skills: identify, communicate

- Provide each child with a $4\frac{1}{4}$" x 14" piece of white paper. Instruct children to fold it in half length wise and then in half again, creating four equal sections. In each section, children should draw and label one part of a food chain so that the strip illustrates a complete food chain.
- Next, children should cut out a circle and several small triangles from the yellow paper to create a Sun. Children can then add details, such as sunglasses.
- Children should then staple the strip to the Sun, placing the Sun before the first drawing in the chain. They can then fold the strip so that it cannot be seen from the front. They can unfold the strip to see the food chain.

Art

Hamburger Food Chain

Materials
construction paper in various colors, $5\frac{1}{2}$" x $8\frac{1}{2}$" pieces of white paper, glue, markers, scissors

Objective: Identify food that comes from plants and animals.

Science Inquiry Skills: identify, communicate, classify

- Discuss with children the food chain for hamburgers: hamburger comes from cows; cows eat hay; hay grows in soil under the Sun.
- Give each child a piece of tan paper and a piece of white paper. They should fold both pieces in half. The tan paper's corners should be cut so that they are rounded to form the hamburger bun. The white paper's corners should be cut and rounded from the folded side. Children should open both papers and glue the white paper to the the tan paper so that the center creases line up.
- On the white paper, children should draw pictures that represent the origin of the contents of a hamburger. For example they can draw a cow and a garden with tomatoes, lettuce, and onions. Then, have children create a hamburger patty and toppings out of construction paper and glue these items to their picture.

Language Arts

Plants and Animals in the Rain Forest

Materials

light-blue, brown, light-green, and dark-green construction paper, glue, markers, scissors

Objective: Identify plants and animals that are found in the rain forest.

Science Inquiry Skills: identify, communicate, classify

- Show children pictures that illustrate the different layers of rain forests. Discuss the types of plants and animals that are found in each layer.
- Give each child a piece of light-blue paper and direct them to fold it in half from top to bottom, then in half from top to bottom again, creating four equal sections. They should open the folded paper.
- Children should use construction paper to create a tree that spans the height of the page and glue it in the center of the page. In each of the four sections, they should draw a layer of the rain forest in the correct order and then label each section with one of the following labels: *emergent, canopy, understory,* and *forest floor*. Children can then decorate the background with animals and other plants and trees found in the rain forest.

Science

Plants and Animals Both Need...

Materials

markers

Objective: Identify shared basic needs of plants and animals.

Science Inquiry Skills: identify, communicate, sort, classify

Resources: "Plants and Animals Both Need..." activity sheet, p. 102

- Lead a discussion about the basic needs of plants and animals. Emphasize needs that are shared, such as the need for light, food, air, water, and shelter. Discuss some of the needs that are not shared, such as plants' need for soil.
- Give each child a copy of the activity sheet. Instruct children to fold the paper in half and then fold up the bottom section of the paper so that the dog and plant look complete. Children should color the pictures.
- In the empty space in the middle of the page, children should write statements about needs that are shared by both plants and animals.

Science

Plants and Animals Need Each Other

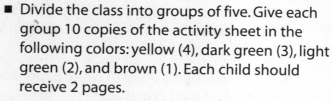

Objective: Identify the interdependence of plants and animals.

Science Inquiry Skills: sort, classify, make a model, compare

Resources: "Plants and Animals Need Each Other" activity sheet, p. 103

- Divide the class into groups of five. Give each group 10 copies of the activity sheet in the following colors: yellow (4), dark green (3), light green (2), and brown (1). Each child should receive 2 pages.
- They should draw the following pictures on each side of their cubes: suns on the yellow cubes; grass, plants, and seeds on the dark-green cubes; plant-eating animals on the light-green cubes; and meat-eating animals on the brown cube.
- Children should construct their cubes by cutting out the cube pattern, folding it along the dotted edges, and taping it together. Each group should build a pyramid by placing the yellow cubes at the bottom, followed by the dark-green cubes, then a layer of light-green cubes, and finally the brown cube on top.
- Have children remove various cubes and discuss what happens. Relate this activity to the interdependence of plants and animals in the food chain. For example, if a sun cube is removed, the whole pyramid collapses. Plants and animals could not survive without the Sun.

Materials

$8\frac{1}{2}$" x 11" pieces of yellow, dark-green, light-green, and brown paper, scissors, crayons, markers, tape

Language Arts

Herbivores, Carnivores, and Omnivores

Objective: Classify animals based on their eating habits.

Science Inquiry Skills: identify, sort, classify, make a model

Resources: "Herbivores, Carnivores, and Omnivores" activity sheets, pp. 104–105

- Discuss the different types of foods eaten by animals, and help children understand the words *herbivores, carnivores,* and *omnivores*.
- Give each child a copy of both activity sheets. Direct children to read each animal's diet on the first activity sheet and decide whether it is an herbivore, a carnivore, or an omnivore. They should write the answers on the lines provided.
- After completing the first sheet, they should list the animals in the correct areas on the second activity sheet and color the diagram.

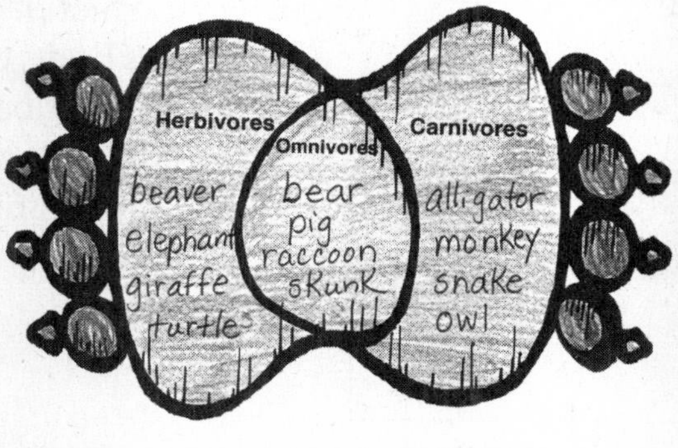

Materials

markers

Science

Insects' Environments

Materials
white paper, colored paints, small round sponges

Objective: Identify various environments in which insects are found.

Science Inquiry Skills: identify, sort, classify, communicate

- Ask children to help you develop a list of insects. Discuss the terms *underground, on the ground,* and *above ground*.
- Give each child a sheet of white paper and direct them to fold it in half and then in half again to create four squares. They should open the paper and draw one of the following in each section: a tree branch with a hive, a leaf, grass, and an underground tunnel.
- Using paint and a round sponge as a stamp, children should paint bees, a caterpillar, an ant, and ladybugs on the correct scenes.

Language Arts

Design an Insect

Materials
markers, construction paper, scissors, stapler

Objective: Identify insect parts.

Science Inquiry Skills: identify, communicate

Resources: "Design an Insect" activity sheets, pp. 106–107

- Give each child a copy of the activity sheets. Ask them to cut the strips along the dotted lines and put them in order. Children should create pages for a booklet by stapling each strip to the bottom of a piece of construction paper. Next, they should staple the pages together to make a booklet.
- On the first page, children should draw the basic shape of an insect's body, then turn the page and draw the new parts of the insect described by each sentence. The final page should show the completed and labeled insect. They should write the name of the insect next to its picture.

Math

Spin an Insect

Materials
markers, brads, paper clips, white paper

Objective: Identify the parts of an insect.

Science Inquiry Skills: identify, communicate, gather and organize data

Resources: "Spin an Insect" activity sheet, p. 108

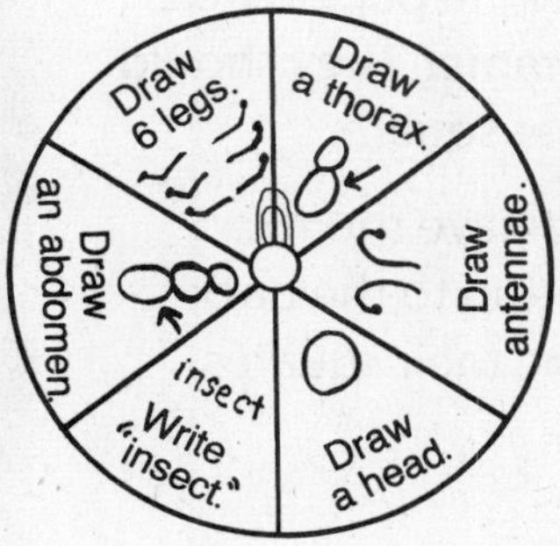

- Give each child a copy of the activity sheet, a brad, and a paper clip. Instruct them to make a spinner by putting the brad through the paper clip and into the center of the spinner on the sheet.
- Give children white paper for sketching. Explain to them how to use the spinner. Their goal is to draw an insect, part by part. For each spin, the child should draw the portion of the insect depicted in the section on which the spinner landed. The parts of the insect should be drawn in order and if the spin lands on a section that is out of order, a tally mark should be made in the box for Insect #1 on the activity sheet.
- After children have completed drawing one insect, they can play again. They should then record their results in the box for Insect #2 and compare it to #1. They should then color their drawings.

Art

Caterpillar to Butterfly

Materials
white paper plates, green and brown construction paper, small beans, pom poms, colored tissue paper, pipe cleaners, packing foam, glue, markers, scissors, small plastic eyes

Objective: Identify the life cycles of insects.

Science Inquiry Skills: classify, make a model

- Guide children in making a list of insects and discussing their life cycles. Focus on the life cycle of a butterfly.
- Give each child a paper plate. Instruct children to divide the plate into four equal sections by drawing two black lines across it so that they make a cross. In each section, a part of the life cycle of a butterfly should be created. Children can use green paper and a bean for the egg on a leaf; green paper, pom poms, and a plastic eye for a caterpillar; brown paper and brown packing foam for the cocoon; and tissue paper and pipe cleaners for the butterfly.

Earth Science

Language Arts — Seasons Quilt

Materials
8″ x 8″ squares of dark-blue and black construction paper, white paper circles, markers, glue, tape, white crayons

Objective: Create a quilt and write about the four seasons.

Science Inquiry Skills: make a model, identify, observe, recognize patterns

Resources: "Seasons Quilt" activity sheet, p. 110

- Provide each child with a square of dark-blue or black paper. Have children draw "stitching" along the edges of the square with white crayons.
- Divide children into groups of four. Each child will be responsible for one season. Have children illustrate a seasonal activity on the white paper circle and glue the circle to a construction paper "frame." They should then glue the frame to the construction-paper square.
- Give children a copy of the activity sheet and have them complete it. They can then glue their description to the back of their "quilt" section. Each group should tape their squares together to create a seasons quilt.

Science — The Months and the Seasons

Materials
12″ x 18″ paper, construction paper in assorted colors, markers, glue, scissors

Objective: Identify the months in each season.

Science Inquiry Skills: identify, collect and organize data

Resources: "The Months and the Seasons" activity sheet, p. 111

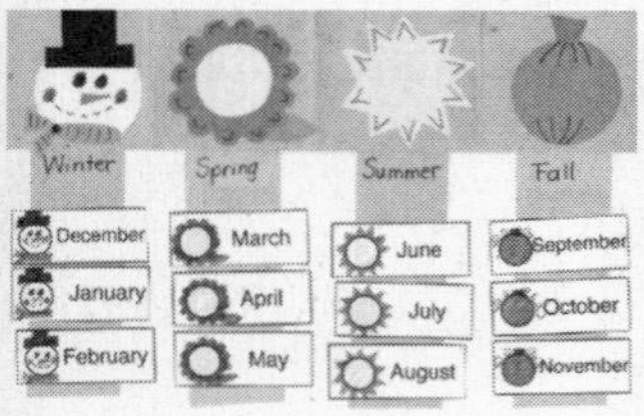

- Give each child a copy of the activity sheet. In the circles next to each month, direct them to draw a snowman for winter months, flowers for spring months, a Sun for summer months, and a pumpkin for fall months. They should then color the symbols and cut out each month's label.
- Give each child a 12″ x 18″ piece of paper and instruct them to fold it in half and then in half again to create four square sections. They should use construction paper to make pictures of a snowman, flower, Sun, and pumpkin, and glue one at the top of each section.
- Children should write the names of the seasons under the appropriate headers. They should then glue the month labels under the correct season.

Language Arts

Seasonal Poetry

Materials
markers, white paper, scissors, stapler

Objective: Describe the four seasons.

Science Inquiry Skills: identify, communicate, compare

Resources: "Seasonal Poetry" activity sheets, pp. 112–113

- Discuss the four seasons with children, and make a list of the colors, smells, activities, weather, and symbols associated with each season.
- Photocopy the activity sheets so that they are on one double-sided sheet, and give a copy to each child. Have children create a booklet by folding the page in half and stapling a cover made from a blank piece of paper along the center crease. Children should illustrate the front cover.
- Ask children to complete the seasonal poems by using the words you discussed or by adding their own.

Science

Seasons Logic Line-Up

Materials
drawing paper, white cardstock, scissors, markers

Objective: Identify the four seasons and put them in order.

Science Inquiry Skills: identify, compare, communicate

Resources: "Seasons Logic Line-Up" activity sheets, pp. 114–115

- Enlarge each seasonal picture from the activity sheet on one piece of white cardstock for use as a class display. Give each child a copy of the first activity sheet and direct them to color and cut out the pictures.
- Read the problems from the second activity sheet out loud and have children rearrange their picture cards to solve them. Then review the correct answers.

Science

Seasons and Our Clothing

Materials

$8\frac{1}{2}$" x 11" white paper, markers

Objective: Identify appropriate clothing for the summer and winter.

Science Inquiry Skills: identify, observe, sort, classify

Resources: "Seasons and Our Clothing" activity sheet, p. 116

- Give each child a copy of the activity sheet and help them identify which clothes are worn in the summer and in the winter. Then discuss differences in seasonal clothing.
- Give each child a piece of white paper and direct them to fold it in half three times so that there are eight sections on the page. They should open the paper, turn it horizontallly, and write "summer" in the center of the left half of the page and "winter" in the center of the right half of the page.
- Children should use the activity sheet to come up with eight outfit combinations for the summer and winter (four for each season). They should draw each outfit in the appropriate section on the paper and note the number and letter of each clothing item above the drawing. For example, a winter outfit could be C3.

Art

Fall Trees Quilt

Materials

8" x 8" squares of paper, brown construction paper, paint in assorted colors, small round sponges, markers, scissors, glue

Objective: Create a seasonal quilt.

Science Inquiry Skills: recognize patterns, observe, compare

Resources: "Fall Trees Quilt" activity sheet, p. 117

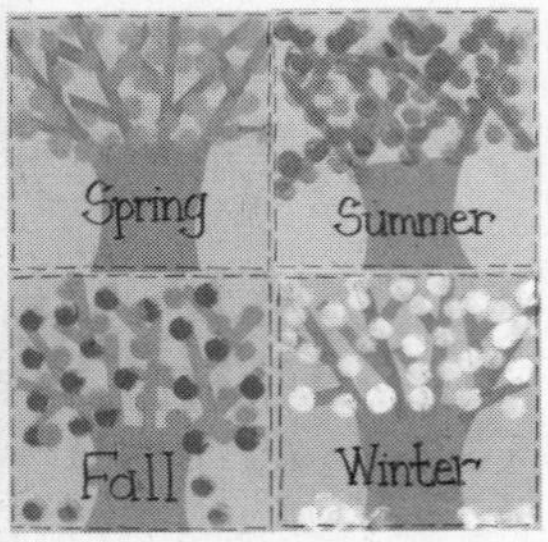

- Divide the class into groups of four. Each child will be responsible for one season. Give each child a square of paper and instruct children to create a tree on it.
- They should use the sponges with colored paints as stamps to create leaves on the trees, choosing appropriate colors for each season. Each square should be labeled with the name of the season.
- Copy the activity sheet for each child. Children should complete the statement for their tree and color the leaf with seasonally appropriate colors.
- Children should create a class quilt by alternating the tree squares with the leaf pages. The seasons should be put in the correct sequence. The quilt can be arranged on a bulletin board or glued onto butcher paper.

Music

Hibernation

Materials

3″ x 12″ construction paper, markers, scissors, stapler

Objective: Identify hibernating animals.

Science Inquiry Skills: identify, communicate, organize data

Resources: "Hibernation" activity sheets, pp. 118–121

- Discuss with children how animals adjust to the cold months. Provide information about hibernating animals and their shelters.
- Put the "Hibernating Animals" song (from the first two activity sheets) on chart paper or provide each child with a copy. Then, teach them the song.
- Provide each child with a copy of the third and fourth activity sheets. Ask children to color the animal pictures, write the animals' names on them, and cut them out. Have children draw and label the hibernation locations in the squares before cutting them out.
- Direct children to staple the animal pictures on the left-hand side of 3″ x 12″ pieces of construction paper and the corresponding location pictures on the right-hand side. Have children write the word "hibernates" in between the two piles.

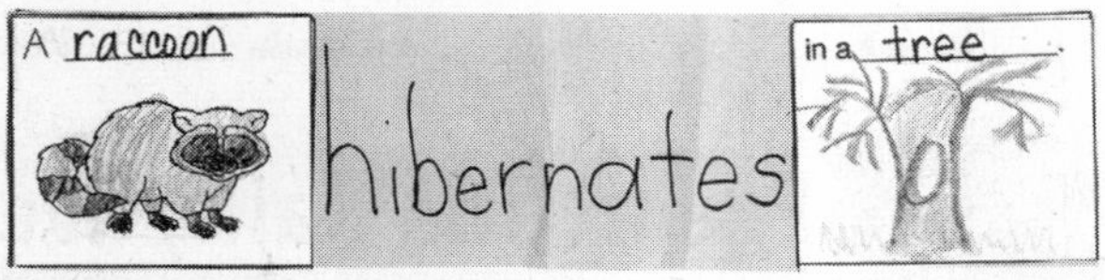

Science

Flip Through the Seasons

Materials

green, red, yellow, white, and pink tissue paper, markers, white paper, glue, 12″ x 18″ construction paper, stapler

Objective: Identify the four seasons.

Science Inquiry Skills: identify, communicate

Resources: "Flip Through the Seasons" activity sheet, p. 122

- Duplicate the activity sheet. Provide each child with four copies. Have children write the names of the seasons in the boxes and add pieces of tissue paper to form leaves or blossoms. On the right-hand side of each tree, children can draw an activity to illustrate each season.
- Ask children to fold each piece of paper in half and cut on the folds. Children can then create a cover page for the book. Next, have children staple the two sides inside a 12″ x 18″ piece of construction paper to create two sides that can be flipped.

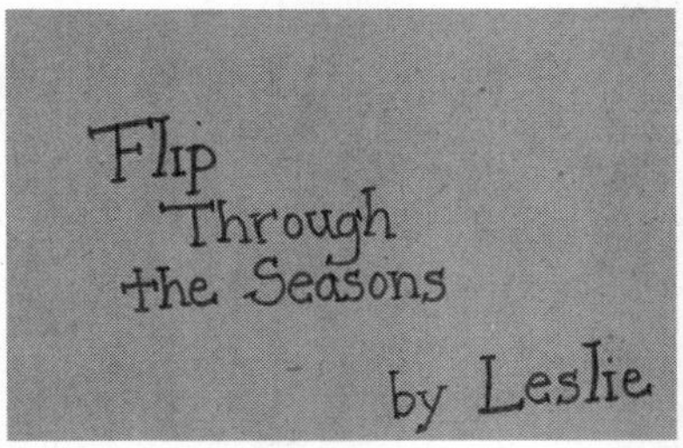

Science

Measuring Weather With Thermometers

Objective: Identify the use of thermometers.

Science Inquiry Skills: communicate, identify, observe

Resources: "Measuring Weather With Thermometers" activity sheets, pp. 123–124

Materials

6″ strips of thin white and red ribbons, tape, scissors

- Duplicate the activity sheets for children and help them read the thermometer next to each picture on the first sheet. Ask children to write the weather words under each picture and record the temperature.
- Assist children in making a paper thermometer out of the second activity sheet. Tape a strip of white ribbon to a strip of red ribbon. Cut slits on the dashed lines at the top and bottom of the thermometer. Then put the taped ribbon through the thermometer to make the gauge.
- Write temperatures on the board and ask children to show the temperatures on their thermometers.

Science

Predicting Rainy Weather

Materials

9″ x 12″ white and gray construction paper, markers, glue

Objective: Identify types of clouds.

Science Inquiry Skills: observe, identify, compare

- Draw different types of clouds on the board and discuss the kinds of weather that can be predicted by looking at clouds.
- Ask children to fold both the white and gray construction paper in half and then in half again to create four sections. Children can then cut the white paper along the folds to make four pieces.
- Have children use the white pieces of paper to create four types of clouds (big, puffy clouds; tall, dark clouds; layers of dark clouds; and wispy clouds). Children should then add color to the clouds and write statements about the clouds. Ask children to glue the left side of each cloud onto the gray paper. Children can lift the clouds and draw the types of weather that the clouds indicate underneath them.
- Types of weather for the clouds can include: "Big, puffy clouds": sun with a few raindrops, nice day but chance of showers; "dark clouds": rain drops, steady drizzle; "tall, dark clouds": heavy rainfall with lightning, thunderstorms; "wispy clouds": sun, fair and sunny.

Science — Measuring Wind

Materials
gallon-size baggies, markers, scissors, strips of construction paper, stapler

Objective: Identify a way to measure wind.

Science Inquiry Skills: observe, identify, communicate

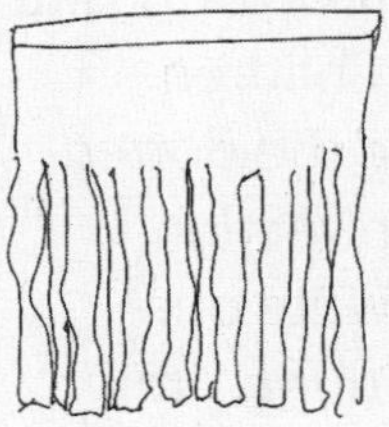

- Tell children that they will create windsocks so they can "see" wind. Provide children with one-gallon baggies and direct them to cut off the bottom. Next, have children cut the baggies from the bottom to roughly one inch from the top, creating many long strips.
- Children can decorate the baggies with markers (these may not remain on the baggies after use) and staple paper strips at the top. Children can then put the windsocks in front of a fan or take them outside on a windy day to watch them move in the wind.

Science — Weather Forecasts

Materials
8" x 14" white paper, scissors, markers

Objective: Predict and observe weather conditions.

Science Inquiry Skills: predict, identify, observe, organize data

- Discuss weather reports and help children identify the following words: *rainy*, *sunny*, *windy*, *snowy*, and *cloudy*.
- Provide each child with a piece of white construction paper and direct them to fold the paper in half from side to side, then in half from top to bottom. Next, have them fold the paper in thirds to create 12 equal boxes. Children should open the paper and cut along the folds on the top section only to create flaps.
- Have children fold the paper from top to bottom. On top of the flaps, ask children to write "(day of week)'s Forecast" for each day of the school week, one in each section, and in the last section, they should write "I predict."
- On Monday, children should draw the type of weather they predict for each day of the week, using a symbol and writing the word to describe the symbol. In the last section, they should predict the number of days for each type of weather.
- Discuss the weather each day with children. Have children lift the flap for that day to draw and write the actual weather for that day.

Language Arts

Weather and Activities

Materials
construction paper, markers, pencils, scissors

Objective: Identify weather conditions that limit activities.

Science Inquiry Skills: communicate, identify, sort, classify

Resources: "Weather and Activities" activity sheets, pp. 125–126

- Discuss with children activities they like to do outside. Ask them what kinds of weather allow them to do outdoor activities and what kinds of weather prevent them from participating in outdoor activities.
- Copy the activity sheets onto one double-sided page so that "I can't" is on the top and "because" is on the bottom, on opposite sides of the page. Provide each child with a copy. Children should fold the activity sheet in half and cut along the dotted lines to create flaps on the top portion. On the top section, children should draw a symbol for weather, then write one statement about an activity that cannot be done during this type of weather.
- On the inside section, have children complete the statement, "I can't ______..." based on the picture they drew. For example, "I can't ride my bike" could be written in a section that says "windy."

Science

Weather and Animal Shelters

Materials
6" x 18" light-blue construction paper, various colors of construction paper, markers, glue, scissors

Objective: Identify shelters that animals use during bad weather.

Science Inquiry Skills: identify, observe, communicate, organize data

- Discuss with children the types of shelters animals use to protect themselves during bad weather. Together develop a list of shelters, for example: *underground*, *in a tree*, *under rocks*, or *in caves*. Children should choose one animal for each location.
- Provide each child with a 6" x 18" piece of light-blue construction paper and ask children to fold the paper in half, then in half again, to create four rectangular sections. Have children write the names of four different shelters and then draw the shelters and an animal for each location.
- Have children add a statement at the top, such as "Weather causes animals to find shelter."

Art

Dressing for Different Weather

Materials
crayons, 9″ x 9″ paper, scissors, glue, markers

Objective: Compare clothes worn in different weather conditions.

Science Inquiry Skills: compare, observe, sort, communicate

- Have children fold a 9″ x 9″ piece of paper into a triangle and then fold it into a triangle again.
- Next, children should unfold the paper and cut along one of the folds to the center of the triangle. Help children place a small amount of glue along the fold to the left or right of the cut and re-fold the paper into a triangle. The triangle shape should now be able to stand upright in a pyramid shape.
- Have children label each section as *rain*, *hot*, and *cold*, and draw clothes in each section that can be worn in those weather conditions. If children wish, they may also cut and paste pictures of clothing from magazines onto their pyramids.

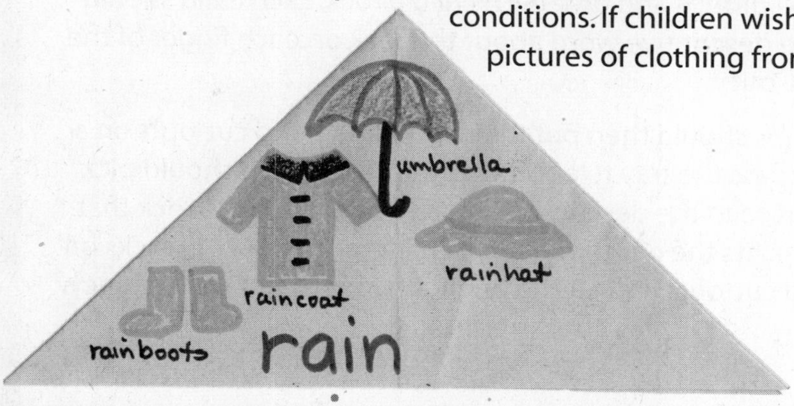

Science

The Influence of Weather

Materials
light-blue construction paper, various colors of construction paper white, glue, markers

Objective: Understand the influence of weather on people and animals.

Science Inquiry Skills: identify, compare, sort

- Discuss with children activities that they like to do in various types of weather and why it is important to have a variety of weather conditions. Emphasize that the Sun controls the weather and also affects animals. Have each child choose one animal that lives in both dry and wet weather. Children should draw, color, and cut out their animals.
- Have children fold a piece of paper in half. On the left-hand side, have children write the label "dry" and draw a scene, adding a Sun to the top, with a statement such as, "The rhino needs the Sun to warm his skin." On the right-hand side, have children write the word "wet" and add a rain cloud with rain, along with a statement such as, "The rhino needs the rain for drinking water." Children should glue their animals to the scenes they have created.

Language Arts — Rocks Are Natural Resources

Materials

collection of rocks, small bag, pencils, trays, scissors

Objective: Identify characteristics of rocks

Science Inquiry Skills: observe, compare, collect and organize data

Resources: "Rocks Are Natural Resources" activity sheet, p. 128

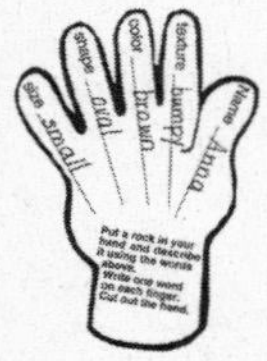

- Put a collection of small rocks in a bag. Read the poem from the activity sheet and let children guess what is in the bag. Help children develop a list of words to describe the rocks based on color, shape, texture, and size, and write the words on the board. Children can add to the rock collection by looking for rocks around the school or near their homes.
- After duplicating the activity sheet for each child, ask children to cut out the hand shape from their sheets. Divide the children into small groups and give each child a rock. Each child should write one descriptive word about the rock on each finger of the hand cut-out.
- The groups should then put their rocks and hand cut-outs on a tray and pass the tray to another group. Each child should pick one hand, read the descriptions, and then choose the rock that they think fits the description. Tell children to leave the rocks on the hand cut-outs and let the original group check for accuracy.

Science — Water Is a Natural Resource

Materials

water, toothpicks, wax paper, eyedropper, tape, markers

Objective: Observe the flow of water in a cycle.

Science Inquiry Skills: observe, predict, compare

Resources: "Water Is a Natural Resource" activity sheet, p. 129

- Provide each child with an activity sheet and have them color the waves, raindrops, and Sun. Discuss the water cycle with children and write the words *precipitation*, *evaporation*, and *condensation* on the board.
- Have children write the words in the correct boxes on the activity sheet. Then they should lay wax paper over the page and tape it.
- Children can then use an eyedropper to place a drop of water somewhere along the water cycle and, using a toothpick, drag the drop through the cycle. Children can repeat the process several times to discover the best way to move the water.

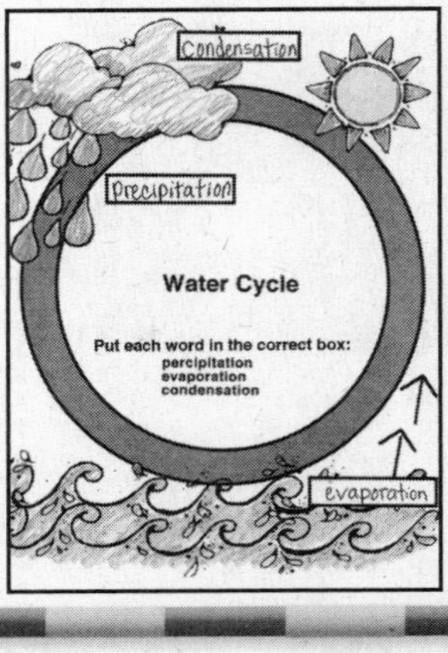

Art

Animals Are Natural Resources

Materials
paper plates, 4" x 8" strips of construction paper, manila, paper, scissors, glue

Objective: Identify a resource chain that involves animals.

Science Inquiry Skills: identify, make a model, organize data

- Discuss with children that animals are natural resources. They are used for work, food, and control of the environment. Ask children to name foods they like and determine which animals they come from.
- Have children create their faces on a circle made from a paper plate. Next, attach a 4" x 18" piece of construction paper behind the "neck" of the face. Children should trace their hands on manila paper, cut them out, and glue them to the ends of the large piece of construction paper. Children should then fold the long strip-like arms toward the center so the hands meet in the front.
- Next, children can design a resource chain involving animals. For example: Sun, grass, cow, milk bottle, and ice cream cone. Have children cut out their illustrations and glue their chains to the hands.

Science

Plants Are Natural Resources

Materials
pencils, markers

Objective: Identify plants in various environments

Science Inquiry Skills: identify, observe, compare, classify

Resources: "Plants Are Natural Resources" activity sheet, p. 130

- Give each child a copy of the activity sheet and ask them to color the locations.
- Have a discussion with children about the types of plants that can be grown in different areas; list some plants and their locations on the board. Help children decide where the plants grow. Children can choose plants from the list and then draw and label them under the correct locations on the activity sheet.

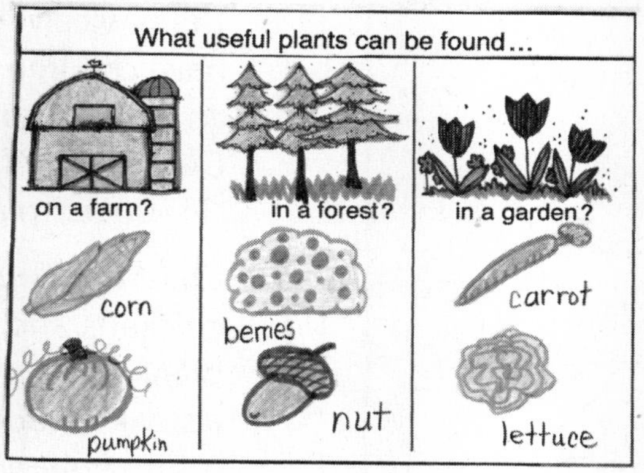

Science

The Importance of Plants

Materials
markers, scissors

Objective: Identify the use of plants as food.

Science Inquiry Skills: observe, identify

Resources: "The Importance of Plants" activity sheets, pp. 131–132

- Help children make a list of fruits and vegetables that are grown in gardens or on farms, and then provide each child with a copy of the activity sheets, copied back-to-back. Children should fold the paper in half so that the words "You Grow" are on the outside. They should then fold the paper in half from side to side two more times to create eight equal boxes. Children should open the paper so that the words "You Grow" are seen, and cut along the fold lines in between the boxes to create four flaps.
- Next, children should draw and label fruit and vegetables from the list, one in each section. Children should then lift the flaps and draw food items that can be made from the fruits and vegetables they have drawn. For example, the top section could state "You grow tomatoes," and a drawing of a tomato could accompany it; the inside could read "to make ketchup" and contain a picture of a bottle of ketchup.

Science

Animals Are Useful

Materials
white paper, crayons

Objective: Identify the uses of animal resources.

Science Inquiry Skills: identify, compare, make a model sort, classify

- Help children develop a list of animals and the products that come from each animal. Sort the items under the following categories: *food*, *clothing*, and *products*.
- Ask children to fold a piece of paper in half twice horizontally, then twice vertically to create 16 boxes. In the top left-hand box, have children write "Uses of Animal Resources" and then draw and label a cow, a chicken, and a sheep across the top boxes. Down the left side, have children write "Food," "Clothing," and "Products," placing one word in each box.
- Children should then draw products that come from cows, chickens, and sheep in the appropriate boxes; some can be left blank. For example, children could draw milk, a hamburger, and a steak as food from a cow; a belt, shoes, and a purse as clothing from a cow; and a leather couch as a product from a cow. The same should be done for the remaining animals.

Science

Wind and Water Resources

Materials
clay, foil, small container with water, straws, paper towel rolls, scissors, markers

Objective: Identify wind and water as resources

Science Inquiry Skills: observe, predict, compare, collect and organize data

Resources: "Wind and Water Resources" activity sheet, p. 133

- Give each child a copy of the activity sheet and ask children to color and cut out the graph.
- Have children build three model boats: one from foil, one from clay, and one from paper. They should then write each boat's material in the first column of their chart. Children should predict which situation will move the boats farthest: by blowing on them through a straw, a paper towel roll, or using only their mouths. They should write their predictions in the middle column of their chart.
- The boats should be tested in a small container of water. After each test, children should record the results in the third column. Children can discuss the results either by telling or writing about them.

	What makes my boat go farthest?	
I made my boat from . . .	Predict	Conclude
foil	straw	
paper	mouth	
clay	tube	

Art

Uses of Land Resources

Materials
scissors, glue, various colors of construction paper, scissors, glue, markers, string

Objective: Identify the uses of soil and rocks.

Science Inquiry Skills: observe, identify, sort, make a model

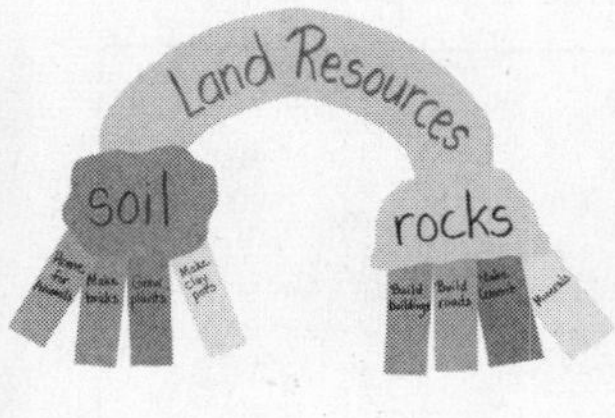

- Help children develop a list of uses for rocks and soil. For example, soil can be used for making bricks and clay pots, growing plants, and as a home for animals; rocks can be used for building homes and roads, making cement, and producing minerals. Children will use this list to create mobiles displaying the products from soil and rocks.
- Have children draw and cut out an arch, a large rock, a mud puddle, and several small strips from colored construction paper. Ask children to label the arch "Land Resources," the mud puddle "soil," and the large rock "rocks." Next, have children write soil and rock products on small strips of paper and attach them to either the rocks or the soil headings. A string can be attached at the top to hang the mobiles.

Language Arts — Fossil Spin and Spell

Materials
markers, scissors, brads, paper clips, pencils

Objective: Identify four types of fossils.

Science Inquiry Skills: compare, observe, identify

Resources: "Fossil Spin and Spell" activity sheets, pp. 134–135

- Discuss the types of fossils found in rocks, shells, and mud, and write the words *shells*, *leaves*, *footprints*, and *insects* on the board.
- Give each child a copy of the activity sheets and ask children to color and cut out the game boards. Have children carefully press a brad through the center of the spinning board and attach a paper clip to the head of the brad to create a spinner.
- Children should spin their spinner and then pick up one letter strip of the word where the spinner lands. They should continue to spin and pick up letters until all of the words have been completed. The words can be built in any order.

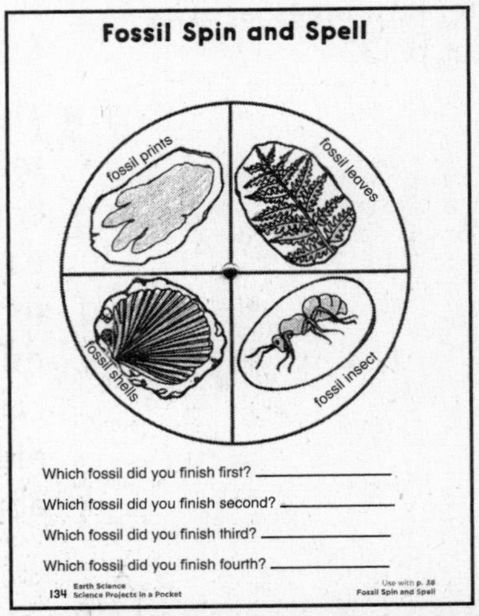

Art — Fossils Are Clues From the Past

Materials
paper bags, paper, markers, scissors

Objective: Understand how scientists study fossils.

Science Inquiry Skills: observe, identify, communicate, collect and interpret data

- Discuss with children dinosaurs and fossils that have been discovered, and show them pictures of these discoveries. Ask children to draw and color pictures of dinosaurs or fossils, making sure that their drawing fills the entire page. Have children cut their drawing into several smaller pieces to create a puzzle and then place the pieces into a paper bag.
- Have children exchange their bags with one another and put together the puzzle pieces in the bag. Next, ask the children to write about the process of discovering the pieces and how they received clues as to what type of dinosaur or fossil they were "discovering" and piecing together.

Art

Fossil Art

Materials

small foam trays, pencils, paints, drawing paper, markers, foam brushes

Objective: Understand how fossils look.

Science Inquiry Skills: identify, observe

- Develop a list of the types of fossils that have been found and draw simple pictures of them on the board. Ask each child to pick one type of fossil. Give each child a foam tray and have them use a pencil to draw the shape of a fossil in the tray. The lines should cut into the foam.
- Next, direct children to brush paint across the trays. Children should lay paper on top of the tray and press down to transfer the design from the tray. They should then carefully lift it off the paper. They may need to try a few prints until they obtain one that they like.
- Allow the paint to dry and then have children add details to the pictures to show the environments in which the fossils were found.

Science

Paleontologists Are Scientists

Materials

modeling clay, small boxes, shells, plastic insects, tree bark, leaves, feathers, popsicle sticks, markers, conctruction paper, yarn, scissors, stapler

Objective: Understand the job of a paleontologist.

Science Inquiry Skills: observe, identify, communicate

Resources: "Paleontologists Are Scientists" activity sheet, p. 136

- Explain to children that a paleontologist is a scientist who studies fossils. Have children color and cut out the visor. Next, children should make a headband by stapling a piece of yarn or a strip of paper to the ends of the visor.
- Have children embed shells, plastic insects, tree bark, leaves, and feathers into clay which has been pressed into a small box. Next, children should cover the box and leave the clay to harden.
- Instruct children to exchange collections. After the clay has hardened, have children turn the boxes upside down and pry the objects out of the clay with popsicle sticks. Children should then record the order in which they found the objects and write statements to describe the fossils. They can wear their visors while they are working.

Math

Dinosaur Footprints

Materials
rulers, paper clips, connecting cubes, books, butcher paper, markers

Objective: Measure a dinosaur footprint in actual size.

Science Inquiry Skills: observe, compare, estimate, predict

Resources: "Dinosaur Footprints" activity sheet, p. 137

- Reproduce the T-Rex footprint from the activity sheet on sheets of butcher paper. Footprints should be 24" x 24," or roughly 4 times larger than the original. This is the same size as the front foot of a T-Rex. Make several footprints.
- Have children use their shoes, rulers, paper clips, connecting cubes, and books to measure the large footprints. Instruct them to record their results on their copies of the activity sheet.

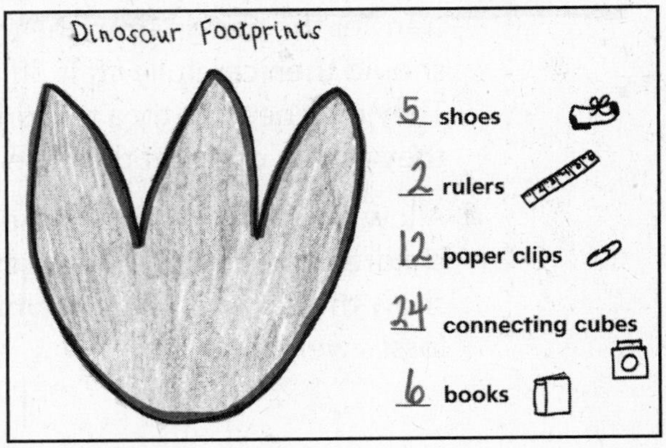

Math

Dinosaur Probability

Materials
brown and green connecting cubes, markers, paper bag

Objective: Complete a probability activity using dinosaurs.

Science Inquiry Skills: estimate, compare, predict, collect data

Resources: "Dinosaur Probability" activity sheet, p. 138

- Give each child a copy of the activity sheet, copied on both sides of one sheet of paper. Have children put 6 green and 2 brown connecting cubes in a paper bag. Children should shake the bags until the cubes are mixed up. Instruct children to pull one cube from the bag, record its color on the activity sheet and return the cube to the bag. Children should do this a total of 10 times. Have children color the circle based on their results.
- Inform children that they will now repeat the experiment. Children should turn their papers over and predict the results before they begin pulling out cubes again. They can write their predictions on the paper and then proceed to repeat the activity as described above.

Science

Dinosaurs

Materials

9" x 12" light-blue construction paper, crayons, markers, glue, scissors

Objective: Identify dinosaurs discovered in the U.S.

Science Inquiry Skills: compare, identify, communicate

Resources: "Dinosaurs" activity sheet, p. 139

- Give each child a copy of the activity sheet. Explain to children that fossils from these dinosaurs are just two of the many dinosaurs discovered in the United States. The two dinosaurs shown on the activity sheet are the Lambeosaurus and the Ankylosaurus.
- Direct children to draw a picture of a landscape in a warm climate on light-blue construction paper. Ask children to leave some space at the bottom of the sheet so they have a place to write later.
- Have children color and cut out the pictures of the dinosaurs and glue them to the landscape. Instruct children to write several sentences at the bottom of their landscape to describe what their dinosaurs are doing.

Art

Dinosaur Creations

Materials

$8\frac{1}{2}$" x 14" white paper, crayons

Objective: Create a dinosaur based on specifications.

Science Inquiry Skills: identify, communicate, organize and interpret data

- Arrange children in groups of four. Have each child fold a piece of $8\frac{1}{2}$" x 14" paper in half twice widthwise to form a booklet. On the cover, children should write "A Dinosaur Creation" along with their names. On the first inside section, instruct each child to draw the head of a dinosaur, creating a plant eater or meat eater, depending on the shape of the teeth. Have them write their names at the bottom of the panel.
- Direct children to pass their papers to the left. In the next section, each child should add on to the dinosaur by creating the front legs and upper body.
- Instruct children to pass the papers to the left again. They should now add the lower body section, including hind legs. After passing the papers once again, children should add a tail in the last section.
- Children should then return the papers to their original owners so that they can study the dinosaurs and write statements based on the new body parts.

Science — Rotating Between Day and Night

Objective: Identify the daytime and nighttime sky in relationship to Earth.

Science Inquiry Skills: identify, classify, compare, make a model

Resources: "Rotating Between Day and Night" activity sheet, p. 141

Materials
light-blue, dark-blue, yellow, and white construction paper, 6" x 9" pieces of black construction paper, stick-on stars, markers, glue, scissors

- Ask children to name items they see in the sky during the day and at night.
- Give each child a piece of light-blue and black construction paper. Direct children to glue the black piece of paper on the right side of the light-blue piece of paper. Then they should cut out and glue a yellow Sun on the blue side and a yellow moon on the black side.
- Children should trace the circle pattern from the activity sheet onto dark-blue construction paper and then cut it out. Instruct children to draw land shapes on the circle so that it looks like Earth.
- Children should glue the Earth to the middle of the paper. They can add pictures to the day and night sides of the paper and write the words "day" and "night" on either side of the paper.

Language Arts — Day Light and Dark Night

Objective: Identify daytime and nighttime activities.

Science Inquiry Skills: compare, sort, communicate

Materials
construction paper in assorted colors, 4" x 4" pieces of paper, $8\frac{1}{2}$" x 11" pieces of paper, 6" x 9" pieces of yellow paper, stick-on stars, markers, scissors, glue, brads, stapler

- Provide each child with a piece of 4" x 4" white paper and direct them to create a circle. Children should then color the top half blue and the bottom half black. They should cut out a Sun, Moon and clouds from construction paper and glue them in the correct section.
- Next, children should fold a white piece of $8\frac{1}{2}$" x 11" paper in half horizontally and then in half vertically to create four sections. They should cut out the four sections and draw 2 daytime pictures and 2 nighttime pictures on them. Then they should write a sentence describing each activity.
- Give each child a 6" x 9" piece of yellow paper for a cover. They should fold it, insert the four pages, and staple everything together. Then they should attach the day and night wheel to the booklet's back cover (on the inside) with a brad so that half of it sticks out. Children should read their books and turn the wheel to match the activity on each page.

Art

The Sun and Shadows

Materials

6″ x 18″ pieces of light-blue paper, 3″ x 4″ pieces of white and black construction paper, markers, scissors, glue

Objective: Illustrate shadows created by the Sun's light being blocked by an object at different times of the day.

Science Inquiry Skills: observe, create a model, compare

- Provide each child with three 3″ x 4″ pieces of white paper and three pieces of black paper and ask them to choose an object that can block the Sun's light. They should draw the object three times on the white paper pieces. Then they should place black paper behind each drawing and staple the two pieces together. Children should cut out the shape of the object from both pieces.
- Give each child a piece of light-blue paper. They should fold it into three 6″ sections. Each drawing of the object should be laid out in a blue section. Then the black shapes should be laid out to illustrate where the shadows fall at different times during the day. Children should draw the sun as rising in the east (right) in the first section to setting in the west (left). The first object's shadow (8:00 A.M.) should be glued to the left of the object. The second object's shadow (12:00 noon) should be glued below the object. The third object's shadow (4:00 P.M.) should be glued to the right side of the object.

Science

Revolving Around the Sun

Materials

black and yellow construction paper, white paper, yellow, red, and orange tissue paper, straws, markers, water, glue, paint brushes or sponges, scissors

Objective: Display the tilt of the Earth as it revolves around the Sun and creates seasons.

Science Inquiry Skills: identify, communicate, classify, compare

Resources: "Revolving Around the Sun" activity sheet, p. 142

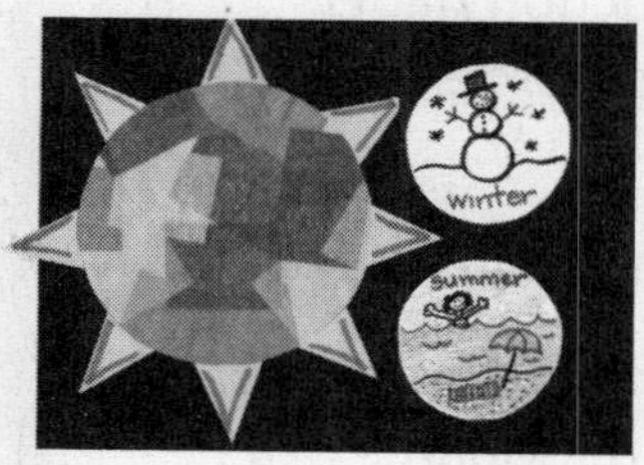

- Lead a discussion about Earth's revolution.
- Provide each child with a copy of the activity sheet and a straw. Children should color the Earth circles and mark their location with a star. They should cut out each circle, place the straw in between the two circles, and glue them together.
- Have children create a Sun by cutting out a large yellow circle and small triangles and gluing the triangles around the circle. Tissue paper can be used to decorate the Sun. It should then be glued on the left side of the black paper.
- Children should illustrate summer and winter on the other two circles and glue them to the right of the Sun.
- Children should use the Earth-on-a-straw to demonstrate the Sun's location in winter and summer, relative to Earth.

Art

The Moon's Surface

Materials

8" x 8" squares of yellow construction paper, pieces of cardboard, glue, sand, markers, crayons, scissors

Objective: Display the Moon's surface.

Science Inquiry Skills: identify, create a model, communicate

Resources: "The Moon's Surface" activity sheet, p. 143

- Brainstorm with children and develop a list of landforms that are found on the Moon, such as craters, mountains, and plains.
- Give each child a piece of cardboard and a copy of the activity sheet. They should cut out and trace the circle pattern onto the cardboard. In the circle on the cardboard, they should use glue to make designs that resemble the Moon's surface. Children should sprinkle sand on top of the wet glue and shake off the extra sand after it is dry.
- When the glue is dry, children should make a rubbing of the moon by laying an 8" x 8" square of yellow paper on top of the cardboard piece and using the side of the crayon to transfer the designs onto the paper. Children should trace the circle pattern onto the yellow square and cut out the shape.
- Children should color and cut out their astronaut, drawing their own face inside the helmet or gluing in a photo of their face. They should complete the statement, "If I landed on the Moon, I ___________," and then glue their astronaut onto the Moon.

Science

Phases of the Moon Quilt

Materials

various colors of paper, 8" x 8" black and purple paper, 6" x 6" black paper, stick-on stars, silver markers, rulers, scissors, glue

Objective: Display four of the Moon's phases.

Science Inquiry Skills: observe, create a model, compare

Resources: "Phases of the Moon" activity sheet, p. 144

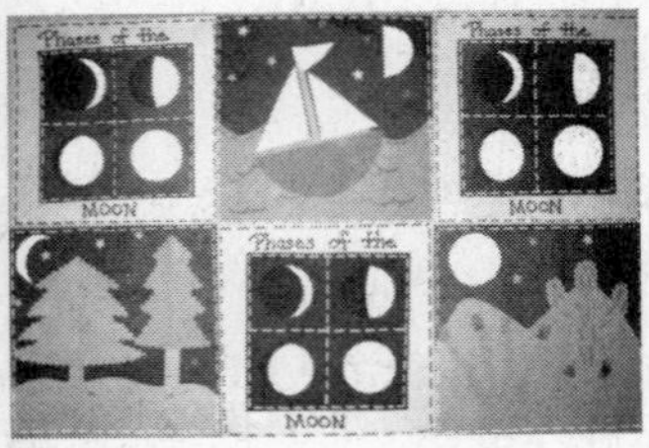

- Discuss with children Earth's rotation around the Sun, and the Moon's movement around Earth. Highlight the phases of the Moon.
- Give each child a copy of the activity sheet and direct them to color the moons yellow, cut them out, and glue them onto 6" x 6" pieces of black paper, each of which has been divided into four equal sections. They should use silver markers to add a stitching border. The black paper should be glued on top of the purple construction paper, entitled "Phases of the Moon," and a stitching border should be added.
- Provide each child with an 8" x 8" piece of black construction paper and direct them to create a nighttime scene.
- Create a classroom "quilt" by stapling children's squares to a bulletin board or gluing them onto butcher paper.

Language Arts

Moon Exploration

Materials
9" x 9" squares of black construction paper, glue, flour, hair spray, markers, scissors

Objective: Learn important dates in space travel to the Moon.

Science Inquiry Skills: identify, communicate, display data

Resources: "Moon Exploration" activity sheet, p.145

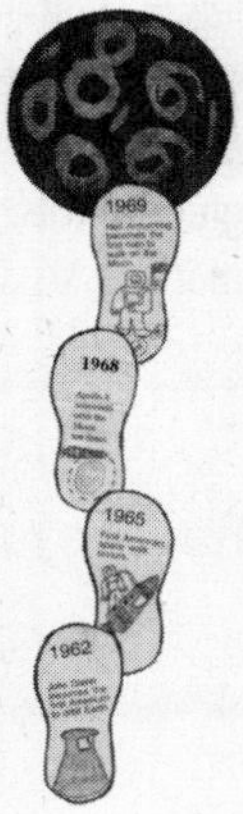

- Read a book about space exploration that focuses on the various space crafts used since space travel began. Discuss John Glenn, Neil Armstrong and other famous astronauts.
- Give each child a black square of paper. Children should cut a circle out of the square by cutting off its corners and rounding off its sharp edges. Next, children should use glue to make random designs on the circle, and then scatter flour on top of the glue. They should shake off the excess flour, ask for assistance with spraying the circle with hair spray, and leave it to dry.
- Give each child a copy of the activity sheet and direct them to cut out the four footprints with statements about space travel to the Moon. Two blank footprints are included for extra events that may be discussed.
- Children should illustrate the events described on the footprints. Then they should glue the footprints to each other, putting the latest event first and moving backwards in time. The chain of footprints should be glued to the Moon.

Math

Many Moons in Space

Materials
markers

Objective: Identify the moons of all the planets in the solar system.

Science Inquiry Skills: identify, communicate, organize, interpret data

Resources: "Many Moons in Space" activity sheets, pp. 146–147

- Display pictures of the eight planets in the solar system and explain that many of the planets have moons. Some have only one moon, and some have many moons.
- Copy the activity sheets back-to-back and give each child a copy. Help children fold the papers so the two sides of the Moon illustration meet in the middle, on the outside of the booklet. Children should color the Moon and stars.
- Children should open the paper and read the directions in the top left-hand corner. Guide children in reading the chart of planets and moons and lead a discussion about the facts in the chart. Children should write a statement in each of the boxes on the sheet and then illustrate their statements.
- Provide assistance to children as they tally the number of moons in the space provided on the right-hand side of the activity sheet.

Music and Art

Planets Go 'Round the Sun

Objective: Identify the planets and learn information about each planet.

Science Inquiry Skills: identify, communicate, make a model, organize data

Resources: "Planets Go 'Round the Sun" activity sheets, pp. 148–149

- Over the course of a few days, children will create models of all of the planets and the Sun. Set up art centers for making models of the planets. Children should complete several projects a day. Use the instructions on the second activity sheet to help children make the planets.
- When children have made all of the models, give each child a copy of the first activity sheet with the song and instruct them to cut out the sections of the song and glue each one to the back of the appropriate planet.
- When all of the models have been completed, give each child $1\frac{1}{2}$ yards of ribbon. Children should glue the top portion of the ribbon to the back of the model of the Sun. They should then glue each planet to the ribbon, following the order that they are mentioned in the song.

Materials

construction paper in assorted colors, $1\frac{1}{2}$" x $4\frac{1}{2}$" strips of orange paper, large piece of paper, paints, paintbrushes, toothbrush, round bath sponge, roller sponges, different sizes of sponges on a stick, bubble wrap, coffee filters, eyedroppers, lids in 5 different sizes, green tissue paper, paper plates, ribbon, glue, pencils, silver markers, scissors

Science

The Solar System

Objective: Identify the order of the eight planets in our solar system.

Science Inquiry Skills: identify, communicate, make a model

Resources: "The Solar System" activity sheet, p. 150

- Display a chart showing the eight planets as they rotate around the Sun. Make a list of the planets, ordered by their distance from the Sun.
- Give each child a strip of blue construction paper and some yellow paper. Direct them to make half of a Sun by cutting a semi-circle and a few small triangles out of yellow paper and then gluing them to the left edge of the blue paper.
- Give each child a copy of the activity sheet. They should create all eight planets by tracing the circle patterns on assorted colors of paper and then writing the first letter of each planet on its circle. They should then glue the planets onto the strip in order, adding rings around Saturn and Uranus with a silver marker. The title "The Solar System" should be written with the silver pen at the top. Children should also complete the mnemonic strip and glue it to the back of the strip.

Materials

6" x 18" piece of dark blue construction paper, various colors of construction paper, markers, scissors, glue, silver markers

Language Arts

Solar-System Riddles

Materials

$4\frac{1}{2}$" x 12" strips of colored paper, $3\frac{1}{2}$" x $8\frac{1}{2}$" strips of colored paper, markers, scissors, glue

Objective: Identify the parts of the solar system.

Science Inquiry Skills: identify, communicate, predict, observe

Resources: "Solar-System Riddles" activity sheet, p. 151

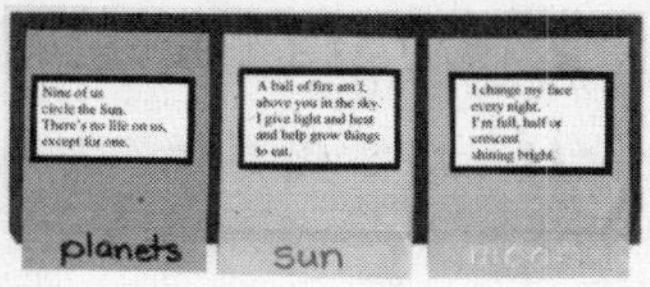

- Read the riddles on the activity sheet to the children and have them identify the answers.
- Give each child a large strip of construction paper and three $3\frac{1}{2}$" x $8\frac{1}{2}$" pieces of paper in assorted colors. Children should turn the small pieces vertically and fold each piece in half horizontally, leaving an extra inch on the bottom. They should then fold the inch of paper up over the folded paper.
- Give each child a copy of the activity sheet. Ask children to cut out the riddles and glue one on the outside of each folded paper on top of the flap. They should open the flap and illustrate the answer and also write the answer on the extra inch at the bottom of the flap. Finally, children should glue the three riddles to their strip of construction paper.

Science

Sorting Planets

Materials

various colors of paper, markers, stapler, index cards, tape, silver markers

Objective: Identify the planets by grouping them into three types.

Science Inquiry Skills: identify, communicate, observe, make a model

- Discuss the order of the planets and show a chart that depicts the Asteroid Belt. On index cards, write the titles "Inner Planets," "Gas Giants," and "Mystery Planets," and tape them onto a wall. Write the names of the planets on index cards and tape them underneath the correct titles. Discuss how the Asteroid Belt separates Mars (the last of the inner planets) and Jupiter.
- Give each child two pieces of construction paper. Children should place the two sheets of paper on top of each other and move the top sheet up about an inch. Then they should fold the papers from top to bottom so that four tabs are created, approximately an inch each. Help children staple the pages together at the top. They should write "Sorting Planets" on the cover and illustrate it with drawings of planets.
- Instruct children to write the titles "Inner Planets," "Gas Giants" and "Mystery Planets" on the tabs. Then they should open each section and illustrate the planets in each category.

Math

Starry Nights

Materials

yellow, blue, and red paper, large piece of white or black paper, rulers, black markers

Objective: Practice counting by 5s and create a number extension chart.

Science Inquiry Skills: observe, make a model

Resources: "Starry Nights" activity sheet, p. 152

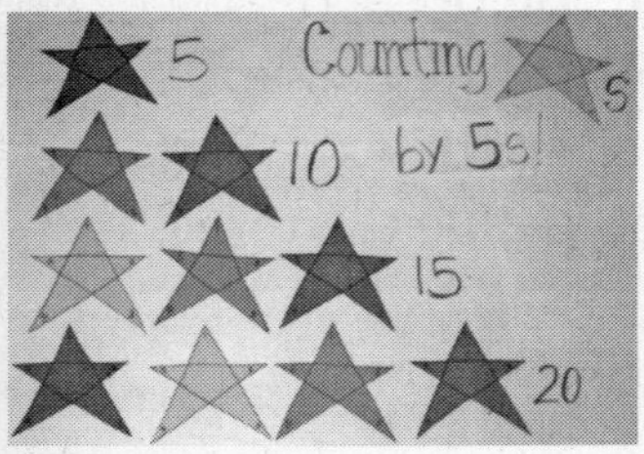

- Lead a discussion about the colors of stars. Tell children that the red dwarf is a star with a very small mass, that the medium-size blue stars are hotter, and yellow stars are cooler.
- Give each child a ruler and a copy of the activity sheet copied on red, yellow, or blue paper. Direct them to draw straight lines connecting the dots between the numbers, in order, and then cut out the patterns. They should write the numbers 1-5 in the five points of the star, beginning at the top and moving clockwise around the star.
- Hang up a large piece of white or black paper and make a "staircase" out of the stars. Start with one star at the top, writing "5" next to it, then two stars in the next row, writing "10" next to them, and so on. Use the same order of colored stars so that a pattern of red, blue, and yellow stars is created.
- The title "Counting Stars by 5s" should be added to the top right of the chart.

Science

Our Sun is a Star

Materials

9″ x 9″ squares of yellow construction paper, purple construction paper, markers, scissors, glue

Objective: Identify information about the Sun.

Science Inquiry Skills: identify, communicate, interpret data, create a model

- Children should trace a hand six times on yellow paper and cut out the hand shapes. Have children write one fact about the Sun on each hand print.
- Give each child a 9″ x 9″ square of yellow paper and direct them to cut off its corners and round off the edges to create a circle. Give each child purple paper and direct them draw a pair of sunglasses. Instruct children to carefully cut out the sunglasses. They should glue the sunglasses on the "face" of the Sun, and the hand shapes should be glued around the edges.

Art

Constellations Galore

Materials

black construction paper, white and yellow paints, paint brushes or spray bottles, stick-on stars, straight pins, potato-chip cans, glue, scissors, can opener

Objective: Identify various constellations in our galaxy.

Science Inquiry Skills: identify, communicate, make a model, observe

Resources: "Constellations Galore" activity sheet, p. 153

- Show children pictures of some well-known constellations. Discuss their names, the number of stars that make up each constellation, and how the star groupings appear.
- Give each child a piece of black construction paper. Children should use paint brushes and spray bottles filled with paint to splatter paint on the paper. When the paint is dry, children can add stick-on stars.
- Give each child a copy of the activity sheet. Direct them to use a straight pin to poke holes through all of the stars to show the constellations, and then to cut out the circles.
- Provide each child with a potato-chip can with the bottom removed, which can be done with a can opener. Children should glue their black paper around the can.
- One circle at a time should be placed in the lid of the can, and the lid should be attached to the can. Children should direct the cans toward light to see the different constellations.

Art

Creating a Constellations Quilt

Materials

8" x 8" squares of black and light-blue paper, yellow paper, silver, red, and black markers, yellow paint, cotton swabs, glue, scissors, scrap paper

Objective: Create new constellations.

Science Inquiry Skills: identify, communicate, make a model, observe

Resources: "Creating a Constellations Quilt" activity sheet, p. 154

- Have children design new constellations by making shapes from dots on scrap paper.
- Give children squares of 8" x 8" black paper and direct them to create new constellations by using cotton swabs and yellow paint to make stars. After the paint dries, they should use a silver marker to connect the dots and name their constellation.
- Give each child a copy of the activity sheet on yellow paper. Children should complete the statements in red marker. Children should cut out their star, glue it to a square of light-blue construction paper, and add a stitching border.
- Staple children's squares on a bulletin board or glue them to butcher paper to create a quilt. Stitching lines should be added to the edges.

Physical Science

Science

The Properties of Matter

Materials
paper bags, pencils, crayons, scissors

Objective: Identify seven properties of matter.

Science Inquiry Skills: identify, compare, communicate, collect and organize data

Resources: "The Properties of Matter" activity sheets, pp. 156–157

- Duplicate the activity sheets for children. Direct children to color the pictures of various types of matter on the first activity sheet and discuss each type of matter.
- Ask each child to choose a different picture. Have children write seven clues about their picture by completing the second activity sheet and cutting out the strips. Direct children to put the strips in a bag. Children should clear their desks of everything except for the picture of the object they described.
- Have children close their bags and put them in the front of the room. Then, have each child randomly choose a bag. They should open the bag and use the clues to search for the picture described.

Language Arts

Properties of Matter Game

Materials
connecting cubes, brads, small paper clips, crayons, scissors

Objective: Identify three types of matter.

Science Inquiry Skills: identify, observe, compare, sort, classify

Resources: "Properties of Matter Game" activity sheet, p. 158

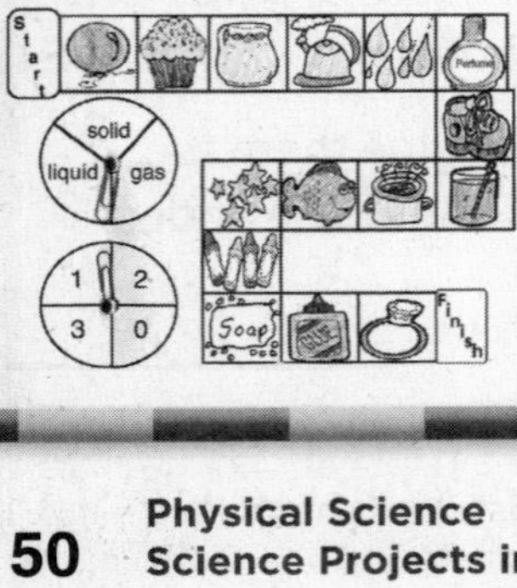

- Distribute one activity sheet to each group of 4 children and assist them in cutting out and assembling two spinners by using brads and paper clips. Children should color the pictures and discuss if the items are solids, liquids, or gases.
- Have each child put a cube at "Start" on their game board and choose a spinner to use. When children spin the matter spinner, instruct them to move their cubes to the nearest example of that matter on the game board. When children spin the number spinner, they should move their cube that many spaces forward and identify the type of matter on which they land. Children can select either spinner each turn. The child who reaches "Finish" first in the group wins the game.

Science

Matter Is All Around

Materials

crayons

Objective: Understand the concept of matter.

Science Inquiry Skills: communicate, identify

Resources: "Matter Is All Around" activity sheets, pp. 159–160

- Give each child a copy of the activity sheets, duplicated on two sides. Direct children to color the pictures.
- Aid children in developing a list of matter items.
- Have children write other examples of that kind of matter in the blanks under each category.

Math

Measuring Matter

Materials

connecting cubes, counting bears, counting tiles, linking loops

Objective: Measure matter in various sizes of jars.

Science Inquiry Skills: observe, compare, estimate, predict

Resources: "Measuring Matter" activity sheet, p. 161

- Give each child a copy of the activity sheet.
- Children should fill one jar with connecting cubes, count the number of cubes used, and estimate the number that will fit in each of the next two jars. Children can then continue the activity by trying different items, making predictions, and checking their results each time. They can then record their results on the activity sheet chart.

Measuring Matter

	Connecting Cubes		Counting Bears		Counting Tiles		Linking Loops	
	Guess	Actual	Guess	Actual	Guess	Actual	Guess	Actual
Jar #1								
Jar #2								
Jar #3								

Science

Heat Changes Matter

Materials

crayons, bread, butter, knife, toaster

Objective: Identify how heat changes the state of matter of food.

Science Inquiry Skills: identify, compare, predict, organize data

Resources: "Heat Changes Matter" activity sheet, p. 162

- Supervise children in toasting two slices of bread. Children should then spread room-temperature butter on the toast and observe the butter change from a solid to liquid.
- Duplicate and distribute the activity sheet to children. Have children think of more foods that go through a change of state when heat is added. Instruct them to pick one such food and draw pictures of what the food looks like before and after heat is added.

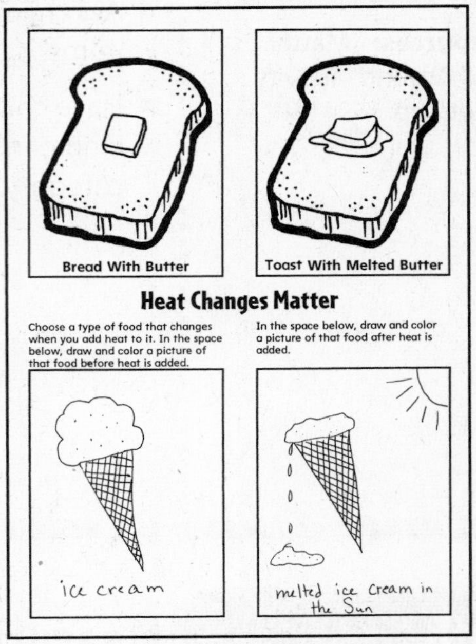

Bread With Butter
Toast With Melted Butter
Heat Changes Matter
Choose a type of food that changes when you add heat to it. In the space below, draw and color a picture of that food before heat is added.
In the space below, draw and color a picture of that food after heat is added.
ice cream
melted ice cream in the Sun

Science

Cooling Changes Matter

Materials

crayons, scissors, pipe cleaners, hole punch, sugar, evaporated milk, vanilla, instant pudding, milk, ice, salt, spoons, small and large ziplock baggies

Objective: Identify how matter can change when it is cooled.

Science Inquiry Skills: identify, compare, communicate

Resources: "Cooling Changes Matter" activity sheet, p. 163

- Give each child a copy of the activity sheet. Direct children to color and cut out the ice cream cone recipe cards and put them in the correct sequence. Children should then punch a hole in the tops of the cards and put a pipe cleaner through the hole to make a recipe booklet.
- Aid children in following the directions on the recipe cards to make ice cream. Discuss how the liquid items, including milk and evaporated milk, come together with other ingredients to form a solid item.

- Children who do not have allergies to the ice-cream ingredients may taste their finished product and describe it.

Science

Matter Magic Hat

Materials
scissors, construction paper

Objective: Identify how matter changes.

Science Inquiry Skills: observe, compare, communicate

Resources: "Matter Magic Hat" activity sheets, pp. 164–165

- Duplicate the activity sheets on construction paper and give one copy to each child. Discuss the four types of changes in matter: solid to liquid, liquid to solid, gas to solid, and liquid to gas. Provide examples to help children recognize these changes.
- Group children in pairs. Have them cut out the strips and the hat, carefully cutting along the two lines on the hat to make slits. Instruct one child in each group to choose a strip and weave it into the hat so that only the pictures and labels can be seen. The other child in each pair should identify the type of change in matter, then pull the strip down to check the answer. Children should take turns with all four strips.

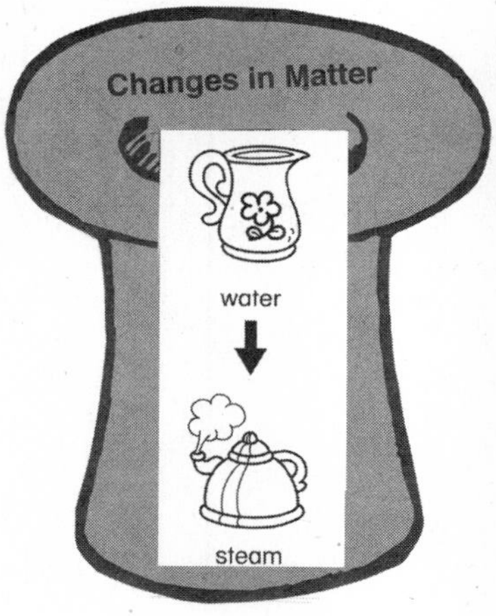

Science

Bowls of Matter

Materials
scissors, crayons, snack baggies, tape

Objective: Understand how some foods have separate pieces of matter and other foods become mixed matter.

Science Inquiry Skills: sort, classify, communicate, observe

Resources: "Bowls of Matter" activity sheets, pp. 166–167

- Explain to children that when we make food, we mix matter. Sometimes we can see the separate ingredients, as in a salad, and sometimes we cannot, as in a cake.
- Give each child a copy of the activity sheets. Have children color the two bowls and cut them out. Then children should tape one snack baggie to the back of each bowl such that the baggie's opening covers the hole in the bowl.
- Direct children to cut out the food strips. They should read each strip and then place it in the appropriate bowl.

Art

Positional Board

Materials

various colors of construction paper, white paper, crayons, scissors, tape

Objective: Identify positional words.

Science Inquiry Skills: identify, observe, compare, communicate

- Discuss with children positional words such as *over, under, on, between,* and *behind*.
- Pair children and have them draw and cut out various outdoor objects such as houses, trees, flagpoles, bridges, and walls. They should leave extra paper at the bottom of each object to fold back and tape to the construction paper so that the objects stand up, creating a 3-D picture. They can then take turns putting the cars or people in different positions and identifying the locations, such as *over the bridge, under the flagpole, behind the wall,* and so on.

Science

Moving Through the Neighborhood

Materials

rulers, crayons, counting bears

Objective: Understand that movement involves changing positions.

Science Inquiry Skills: observe, identify, communicate

Resources: "Moving Through the Neighborhood" activity sheets, pp. 168–169

- Distribute the first activity sheet, which contains a map of The 3 Bears' neighborhood. Direct children to color the pictures on the map.
- Next, give each child some counting bears and a copy of the second activity sheet. Have children read each description of movement on the sheet. Direct children to follow the movements of the bears, then use a ruler to measure the distances between their positions. Children should record the measurements on the sheet.

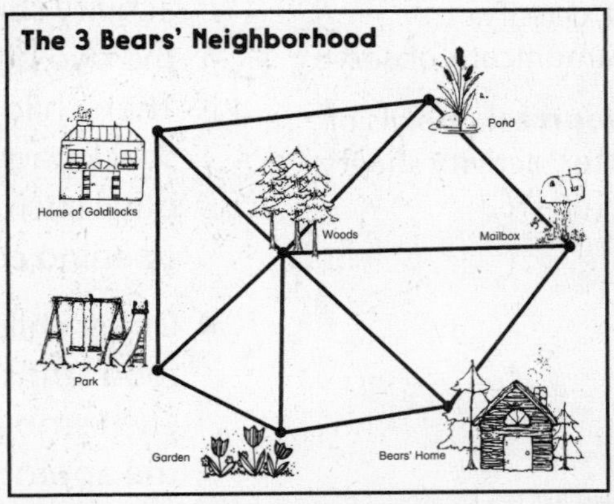

Science

Speed and Friction

Materials

crayons, scissors, stapler, bubble wrap, aluminum foil, sandpaper, carpet, wooden blocks, small toy cars

Objective: Understand that speed is affected by the friction of surfaces.

Science Inquiry Skills: observe, compare, predict

Resources: "Speed and Friction" activity sheet, p. 170

- Provide each child with a copy of the activity sheet. Ask children to color and cut out the car and the strips below the car. Instruct them to stack the strips, face up, on the car and staple them at the top.
- Distribute bubble wrap, aluminum foil, sand paper, carpet, and toy cars to children. Give them wooden blocks to create a variety of ramps.
- Have children roll toy cars down each surface and record the results on the strips.

I rolled my car across bubble wrap
My car went "pop".

Science

Push-or-Pull Blackout

Materials

pencils, pennies

Objective: Identify objects that are pushed or pulled.

Science Inquiry Skills: observe, compare, identify

Resources: "Push-or-Pull Blackout" activity sheet, p. 171

- Assist children in making a list of items that are pushed and items that are pulled, and discuss the different movements.
- Give each child a copy of the activity sheet and 10 pennies. Direct them to fill the squares with the words listed on the top of the page. Children should flip the pennies and cover the squares based on whether the item in the square is pushed (heads) or pulled (tails). All items must be covered to achieve blackout.

Push-or-Pull Blackout
swing
trailer
cart
wagon
Head-push
Tail-Pull
Free
stroller
mower
sled
water-skier

Language Arts

Machine Match-Up

Materials
scissors, crayons

Objective: Identify four types of machines.

Science Inquiry Skills: compare, sort, classify, communicate

Resources: "Machine Match-Up" activity sheets, pp. 172–173

- Discuss with children ways in which simple machines help people do jobs, focusing on the following: wheel and axle, wedge, lever, and inclined plane.
- Provide children with the activity sheets and direct them to color and cut out the cards. Group children in pairs and give them the choice of playing "Memory" with one set of cards or "Go Fish" with two sets.

Science

Gravity Bear

Materials
crayons, scissors, pennies, tape

Objective: Recognize that gravity helps balance.

Science Inquiry Skills: identify, compare, observe

Resources: "Gravity Bear" activity sheet, p. 174

- Give each child a copy of the activity sheet. Instruct children to color and cut out their bears. Next, have them tape pennies behind the bear's front paws. Have children read the short statement about gravity.
- Children can balance the bear by placing a finger under the head and folding the bear's chin under it slightly.

Science — Magnetic Forces

Materials

crayons, scissors, magnets, items that are attracted by magnets and items that are not attracted

Objective: Identify objects that are attracted by magnets.

Science Inquiry Skills: identify, compare, sort, classify, predict

Resources: "Magnetic Forces" activity sheets, pp. 175–176

- Give each child a copy of the activity sheets, including two copies of the second activity sheet. Instruct children to cut along the dotted lines on each sheet. They should then staple the pages together to make a book.
- Provide children with a variety of objects that they can test for magnetic attraction. Suggested items include paper clips, safety pins, scissors, screwdrivers, pennies, nails, food, pencils, erasers, crayons, and small wood blocks.
- Direct children to test the items with magnets and record the results in their booklet by filling in the blanks. They can draw and color the object in the blank space.

Science — Loud and Soft Sounds

Materials

scissors, film canisters, toothpicks, rice, number cubes, pennies, paper clips, pebbles, pom poms, marbles

Objective: Identify loud and soft sounds.

Science Inquiry Skills: compare, identify

Resources: "Loud and Soft Sounds" activity sheets, pp. 177–178

Loud Sounds	Soft Sounds

- Duplicate the activity sheets on two-sided paper, with the 16 unlabeled canisters on the left-hand side of the back of the paper.
- Fill 8 small film canisters with the items from the materials list and label the canisters #1–#8. Next, fill 8 more canisters with the same items and label them A–H.
- Direct children to fold the sheet in half and then turn to the side with the 16 labeled canisters. Next, children should shake the film canisters. On the blank line next to each number, have children write the letter of the matching canister. When children are finished, they can open the canisters to check their answers.
- Have children turn the paper over and cut along the dotted line, creating two flaps. Children can open the Loud Sounds flap and write the numbers and letters of canisters that make loud sounds. Have them do the same with the Soft Sounds flap.

Science

Sources of Heat

Materials
6″ x 18″ construction paper, white, blue, yellow, and manila paper, markers, crayons, scissors, glue

Objective: Identify four sources of heat.

Science Inquiry Skills: identify, observe, create a model

- Discuss various sources of heat, such as the Sun, electricity, friction, and Earth's interior, and make a list on the board.
- Provide each child with a 6″ x 18″ piece of construction paper, which they should fold twice lengthwise to create four equal sections. Also, give children squares of yellow, white, manila, and blue paper. Children should cut a circle from the yellow paper and draw several uses of the Sun's heat, such as a solar panel and a greenhouse, on the circle.
- Next, have children draw a light bulb and illustrate two forms of electricity on the bulb. Children should then cut a circle from the blue paper, adding land shapes and two forms of Earth's heat, such as volcanoes, on the circle to depict Earth.
- Ask children to trace their hands on the manila paper and cut them out. Then they should draw several methods of friction, such as tires braking, on the hand.
- Children should glue the pictures to the blue paper, one in each section, and add the labels Sun, electricity, friction, and Earth's interior.

Art

The Sun Warms Everything

Materials
yellow, orange, and white paper, markers, crayons, scissors, brads, glue

Objective: Display parts of Earth that the Sun heats.

Science Inquiry Skills: identify, observe, create a model

Resources: "The Sun Warms Everything" activity sheet, p. 180

- Discuss with children how Earth helps warm the land, air, and water.
- Duplicate the activity sheet on white and yellow paper for each child. Direct them to cut out the circles and a $\frac{1}{3}$ section of the yellow circle and write "What does the Sun warm?" on the remaining yellow section. Then have them trace the edges of the yellow section with an orange marker.
- Next, children should divide the white circle into three equal pie-shaped sections. They should color and label three scenes, one in each section, depicting air, land, and water. Children should cut small triangles from the orange paper and glue them to the back of the white paper to create the Sun's rays.
- Have children place the yellow section on top of the white circle and connect the two circles with a brad through the middle. The top circle can be turned to show one $\frac{1}{3}$ section at a time.

Language Arts

Changing Temperatures

Materials
white paper, crayons, markers, scissors

Objective: Identify various degrees of heat and symbols for the degrees.

Science Inquiry Skills: compare, identify, sort

Resources: "Changing Temperatures" activity sheet, p. 181

- Lead a discussion about temperature and, as a class, develop pictures to illustrate four different temperatures: 0, 32, 68, and 90 degrees Fahrenheit.
- Provide children with white paper and direct them to fold the paper in half from top to bottom, then again from side to side, to create four equal sections. Label each section with one of the following: 0, 32, 68, or 90 degrees Fahrenheit.
- Children should draw the pictures that the class developed, one in each square, to correspond to the temperature. Then ask children to cut the paper into four cards.
- Read riddle 1 aloud and ask children to place their cards in order on their desktop according to the riddle. All four cards should be used in each riddle. Continue with riddles 2–4.

Art

Containing Body Heat

Materials
manila construction paper, various colors of construction paper, markers, crayons, glue, scissors

Objective: Identify clothing we use to contain body heat.

Science Inquiry Skills: identify, communicate, organize data, create a model

Resources: "Containing Body Heat" activity sheets, pp. 182–183

- Explain to children that heat can be contained. Use examples such as: a lid holds heat in a pan, and an oven door contains the oven's heat. Discuss the idea that we wear certain clothes to contain our body heat in cold weather.
- Give each child a copy of the first activity sheet. Read aloud the sentences on the paper with your class. Children should complete each sentence by writing the correct type of clothing on each line.
- Next, provide each child with a copy of the second activity sheet and direct children to cut out the paper doll. They should trace around the hands on colored paper to make mittens, around the top part of the body for a jacket, around the feet for shoes, around the legs for pants, and around the head to create a hat. Then each child should "dress" their paper doll, adding details to the clothing if they wish.

Science

Sounds in Our World

Materials
scissors, glue, magazines

Objective: Classify environmental sounds as nature or manmade sounds.

Science Inquiry Skills: identify, sort, classify

Resources: "Sounds in Our World" activity sheets, pp. 184–185

- Ask children to tell about environmental sounds that they hear daily. Make a list on the board under the labels "Nature Sounds" and "Manmade Sounds," helping children decide in which category each sound belongs.
- Provide children with the first activity sheet, which shows pictures and sounds. Direct children to cut out the pictures and place them, without glue, on the sorting board from the second activity sheet. Each child should work with a partner and check if their answers match. They can discuss any differences they had in their sorting. Children can also play "Go Fish" in groups of four by combining all of their cards.
- Children can then use the sorting mats as a place to glue magazine pictures that depict nature sounds and manmade sounds.

Language Arts

Environmental Sounds Card Game

Materials
scissors, crayons, markers

Objective: Match various sounds to the objects which make those sounds.

Science Inquiry Skills: identify, communicate, sort, classify

Resources: "Environmental Sounds Card Game" activity sheets, pp. 186–187

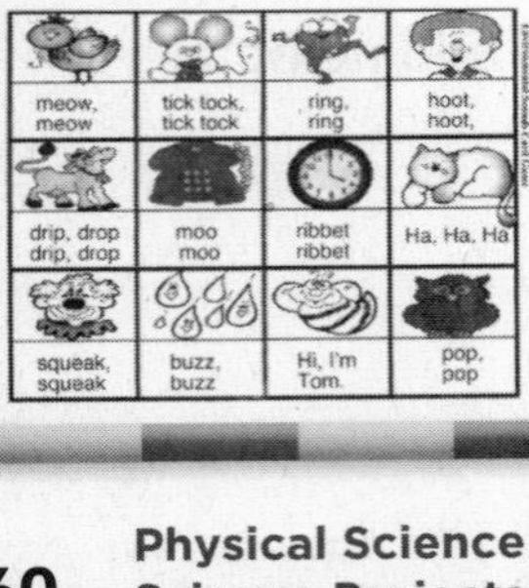

- Duplicate the activity sheets. One set of cards can be enlarged and used with the entire class, or small sets can be used in groups to play the game.
- Pass out one card to each child. Ask one child to start by making the sound on the bottom portion of his or her card. The child who has the card with the object that matches the sound should identify the picture. For example, if "ring, ring" is said, then the child with the telephone picture should say "telephone." The child who identified the picture should then say the sound on the bottom of his or her card. This continues until all of the sounds have been matched. This game can also be played in smaller groups.
- Each child can color one set of cards each and take them home to play with family members and playmates.

Art

Loud and Soft Sounds

Materials

9" x 12" gray construction paper, various colors of construction paper, markers, crayons, scissors, glue, brads

Objective: Identify ranges of loud and soft sounds made in nature and in our environment.

Science Inquiry Skills: identify, sort, organize data, create a model

Resources: "Loud and Soft Sounds" activity sheets, pp. 188–189

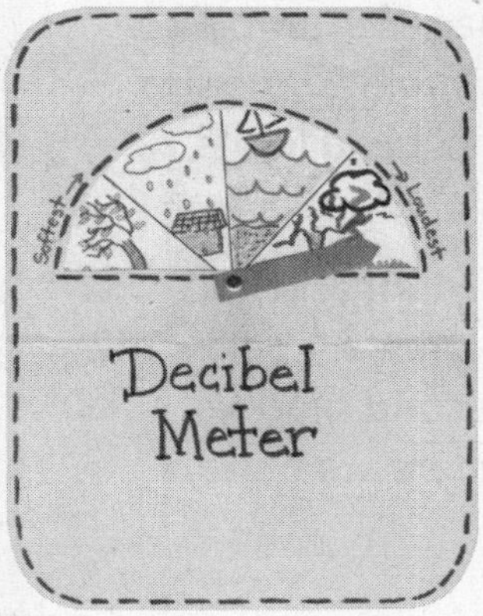

- Lead a discussion about sounds in nature. Make a list of the sounds in our environment made by objects, and rate them from soft to loud.
- Children should cut a small triangle off of each corner of a gray piece of 9" x 12" paper and round off the sharp edges. Have children trace the semi-circle from the activity sheet onto the top half of the "Decibel Meter" and cut it out to create a window.
- Provide each child with one colored circle and $\frac{1}{2}$ circle of another color. Children should glue the semi-circle on top of the complete circle and divide it into eight sections. Children should draw one nature sound on each of the four sections of one color, in the order from softest to loudest sounds. On the other side of the circle, children should draw four sounds that objects make in our environment, in order from softest to loudest.
- Children should trace the arrow from the activity sheet on red paper and cut it out. They should place a brad through the arrow, the "Decibel Meter," and the circle to connect all three. Children should write the labels *softest*, *loudest*, and *"Decibel Meter."*

Science

Conducting Sounds

Materials

coins, markers, scissors, stapler, tape

Objective: Identify items as good or poor conductors of sounds.

Science Inquiry Skills: identify, communicate, organize data, classify

Resources: "Conducting Sounds" activity sheets, pp. 190–191

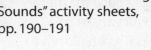

- Explain to children that some items, such as a table, are good conductors of sound, and other items, such as a soft towel, are poor conductors of sound.
- Duplicate the activity sheets and provide coins for children to drop onto items. Have children cut along the dotted lines on both sheets to create six strips and one header. Layer the strips and staple them together.
- Next, each child should drop a coin onto various surfaces around the classroom. The surfaces should be recorded under "Good Sound Conductors" or "Poor Sound Conductors," depending on the level of sound produced.
- Children should then write the name of the item on the dotted line of the strip, draw the item, and circle either "good" or "poor" in each sentence. The strips should be stapled to the half sheet, and a coin can be taped over the drawing.

Art

Sources of Light

Materials
yellow copy paper, white paper, markers, crayons, scissors

Objective: Identify natural and artificial sources of light.

Science Inquiry Skills: identify, observe, sort, communicate

Resources: "Sources of Light" activity sheet, p. 192

- Lead a discussion about sources of light and make two lists: "Natural" and "Artificial." Provide each child with a copy of the activity sheet, duplicated on yellow paper.
- Direct children to cut out the light bulb and staple it to the top of a stack of four pieces of white paper. Ask children to cut along the edges of the yellow light bulb to make the white papers into the same shape.
- Ask children to draw two forms of natural light and two forms of artificial light, one on each piece of white paper. Examples of natural light are: Sun, stars, lightning, meteors, fireflies, and fire. Examples of artificial light are: a flashlight, lamp, head lights, street lights, candle, and neon signs.
- Children should draw four different sources of light and add a statement to each picture. First graders can write a label for each drawing, such as "Moon and Stars." Second graders can write a sentence such as, "The Moon and stars are natural sources of light."

Art

Light Travels in a Straight Line

Materials
9" x 12" gray construction paper, 6" x 18" yellow construction paper, markers, crayons, scissors

Objective: Describe light sources and the concept that light travels in straight lines.

Science Inquiry Skills: identify, observe

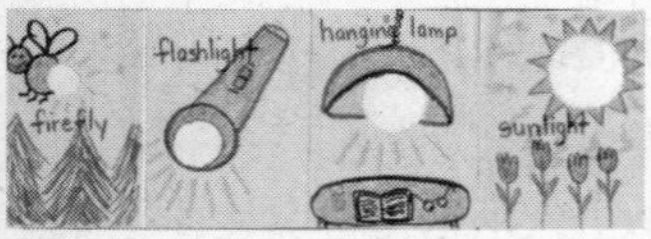

- Encourage children to brainstorm and list sources of light in our environment. Review that light travels in straight lines.
- Direct children to fold a 9" x 12" piece of gray construction paper in half and then in half again to create four equal boxes. Ask them to draw four items that create light and then cut a circle out of each drawing to provide space for the light to shine through. The sizes of the circles will vary from a small light for a firefly to a lot of light for the Sun. Ask children to cut along the folds of the gray paper to create four cards.
- Next, children should fold a 6" x 18" piece of yellow paper into four sections and glue the drawings, one in each section, to create four pictures of light. Children can draw light rays in straight lines coming from each light with yellow markers, and then label each source of light.

Art

Light Can Be Blocked

Materials

$8\frac{1}{2}$" x 11" blue paper, various colors of construction paper, crayons, markers, scissors

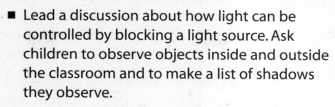

Objective: Understand shadows are created by blocked light.

Science Inquiry Skills: identify, sort, communicate

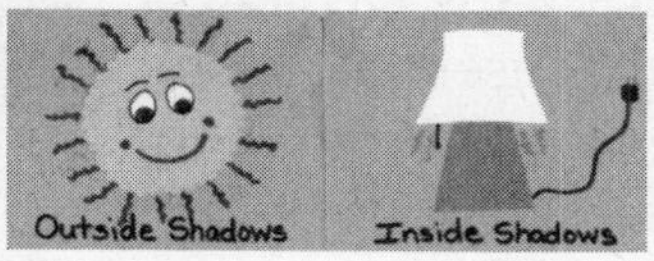

- Lead a discussion about how light can be controlled by blocking a light source. Ask children to observe objects inside and outside the classroom and to make a list of shadows they observe.
- Give each child a piece of blue construction paper and instruct them to fold the paper in half from top to bottom, then again from side to side. Children should reopen the paper to the first fold and cut along the fold to create two flaps. Ask children to use the colored construction paper to create a Sun on one side and an object that creates inside shadows on the other side. Have children label the respective sides "Outside Shadows" and "Inside Shadows."
- Next, children should open the flap under the Sun and draw items from their list that have shadows formed by an object that is blocking the Sun's light. Children should then open the other section and draw objects and their shadows that are formed by an inside light source being blocked. They should also label the objects.

Science

Light Spectrum

Materials

white paper, crayons, large paper clips, tape, scissors

Objective: Understand that white light is a mix of all colors and that the colors form a band called a spectrum.

Science Inquiry Skills: identify, sort, make a model

Resources: "Light Spectrum" activity sheet, p. 193

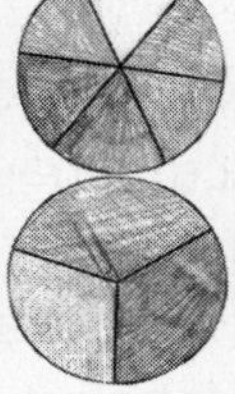

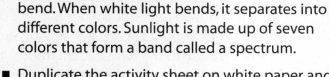

- Explain that color is seen because light can bend. When white light bends, it separates into different colors. Sunlight is made up of seven colors that form a band called a spectrum.
- Duplicate the activity sheet on white paper and direct children to color the circle with three sections red, blue and green. Children should color the six-section circle red, orange, yellow, green, blue, and violet. Then they should cut out both circles.
- Direct children to straighten one side of a paper clip, poke it through the center of each circle, and use it as a spinner.
- Children should spin the three-color circle to see the three colors blend together. Next, they should spin the six-color circle and watch the colors blend together into a bright, creamy color. White light could be seen with a seven-section circle that includes the color indigo.

Art

Electricity in Our Daily Lives

Materials

$8\frac{1}{2}$" x 11" white paper, various colors of construction paper, crayons, markers, stapler, scissors

Objective: Identify ways which electricity is used in daily life.

Science Inquiry Skills: identify, observe, communicate

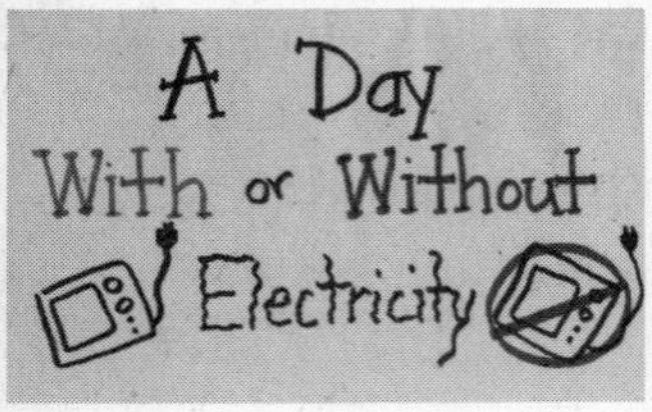

- Help the class develop a list of the ways we use electricity in daily life, and discuss how each activity could be done without electricity.
- Provide children with colored paper, and instruct children to fold the paper in half from top to bottom to make the cover. Children should write "A Day With or Without Electricity" on the cover and draw a picture of an object that uses electricity, twice, with a red circle and slash over one of the images.
- Next, children should fold two pieces of white paper in half from top to bottom, and cut up the middle to form two sections, which should then be stapled inside the cover. On the left side of each paper, children should draw an activity that uses electricity and write the sentence, "With electricity, I would …" On the right sides, children can draw how they would do the same activity without electricity and write, "Without electricity, I would …"

Science

Current Electricity

Materials

7" x 9" brown paper, 6" x 8" gray paper, various colors of construction paper, white paper, markers, glue, scissors

Objective: Understand how electricity travels through a circuit.

Science Inquiry Skills: identify, observe, make a model, communicate

- Lead a discussion about items that use electricity in our environment. Draw a circuit path with a generator, electrical wires, a light switch, lamp, and the return of the electricity.
- Children should make a TV by using a 7" x 9" piece of brown paper and a 6" x 8" piece of gray construction paper. They should cut a 5" x 5" section out of the gray paper, justified close to the left side, and then glue the gray paper on top of the brown paper along the top and bottom edges only. Children should also create an antenna and buttons.
- Give children white paper. Direct them to fold the paper in half lengthwise and then cut down the middle along the fold. The two pieces should be glued together to form a long strip that should then be folded into 8 sections.
- In the first section, children can write "Electricity Travels …" and proceed to create the following pictures with construction paper: 2 generators, brown poles, a house, a light switch, and a lamp. They should glue one picture to each section, placing a generator at the beginning and the end. Each section should be labeled.

Science — Static Electricity

Materials
white paper, colored tissue paper, balloon, scissors, glue, markers

Objective: Identify static electricity.

Science Inquiry Skills: observe, make a model

Resources: "Static Electricity" activity sheet, p. 194

- Explain to children that an object with a static electrical charge acts like a magnet. This activity will work best when the temperature is cold and the humidity is low.
- Provide each student with a copy of the activity sheet, and direct children to cut out the pattern of the butterfly wing. They should lay the pattern on top of a piece of white paper, which has been folded in half from top to bottom, placing the dotted line of the pattern along the fold. They should cut out the pattern and trace it onto tissue paper to create another butterfly. They should also cut out the body from the activity sheet and color it. Then ask children to layer the butterflies, placing the tissue paper on top and gluing the body in the middle.
- Children should inflate and tie their balloons, and then charge them by rubbing them on their hair. They should hold the charged balloon near the wings and then move it away, causing the wings to "fly" up into the air. Explain how the balloon gathers electrons off of their hair, making the balloon negatively charged. The reaction between the negatively charged balloon and the positively charged tissue causes the wings to "fly."
- Children can also experiment by charging the balloon and holding it near packing foam pieces or their hair.

Language Arts — Inventors and Electricity

Materials
scissors, crayons, markers

Objective: Identify four inventors involved with electricity.

Science Inquiry Skills: identify, communicate

Resources: "Inventors and Electricity" activity sheet, p. 195

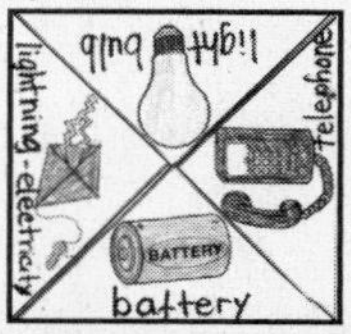

- Read stories to children about inventors who influenced the development of electric inventions. The inventors discussed in this activity are Benjamin Franklin, Alexander Graham Bell, Thomas Edison, and Alessandro Volta.
- Give children the activity sheet, and read the information in the middle of the paper. Next, tell children to cut out the square and fold each corner into the middle so that the points meet.
- Children should color and label each picture and write the name of the inventor below the correct invention. Children can also write a sentence about each inventor.

Activity Resources

The Parts of Plants

The Needs of Plants

Plant Environments

Plants Grow and Change

Which plant part do you like best?			
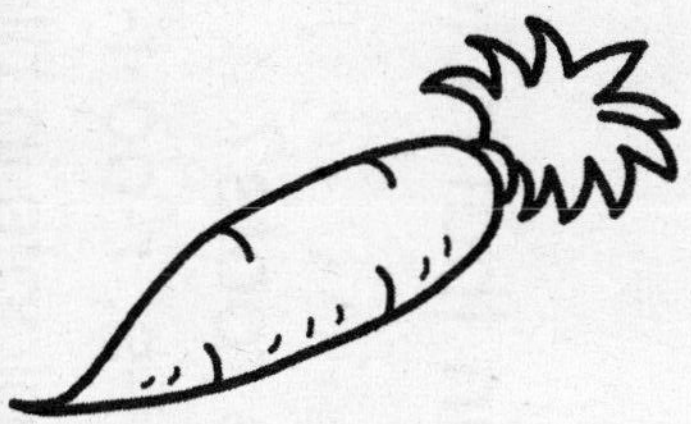 root	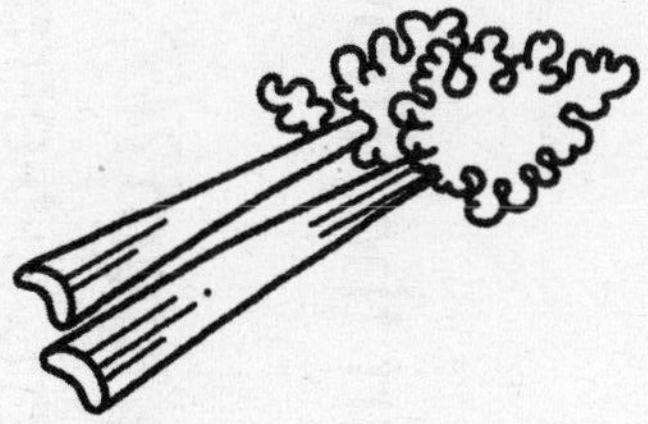 stem	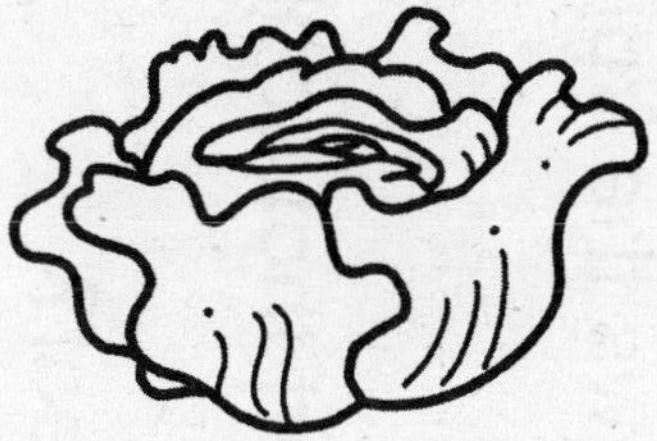 leaves	 flower

What Is the Part?

(tune: BINGO)

What is the part that grows
the fruit?
Could it be the flower?
Fl-ow-er, fl-ow-er, fl-ow-er,
The flower grows the fruit.

What is the part that soaks
up light?
Could it be a leaf?
L-e-a-f, l-e-a-f, l-e-a-f,
The leaf soaks up the light.

What is the part that moves
water up?
Could it be the stem?
S-t-e-m, s-t-e-m, s-t-e-m,
The stem moves the water up.

What is the part that holds
the plant?
Could it be the roots?
R-oo-t-s, r-oo-t-s, r-oo-t-s,
Roots hold the plants in place.

Two-Inch Quilt Squares

Use with p. 3
Flower Parts Quilt

Flower Glyph Legend

1. Do you water plants at home?
   Yes: Add light-blue rain.
   No: Add dark-blue rain.

2. Where do you have plants?
   Outside only: Add dark-brown soil.
   Inside and outside: Add light-brown soil.

3. Have you ever planted anything in your yard?
   Yes: Bend the flower to the right.
   No: Bend the flower to the left.

4. Do you have a garden?
   Yes: Add a sun.
   No: Add a cloud.

Use with p. 4
Flower Glyph

A pine tree has leaves that stay green in cold places.

Plants Grow Everywhere

by ________

wet

Plants have special parts to grow in different places.

A cactus has spines to grow in hot, dry places.

cold

Plants Grow Everywhere

by ____________

A cattail has roots to grow in wet places.

Plant Cube Toss

p	l	a	n	t	s

Logic Line-up: How Fruit Grows on a Tree

Problem 1

1. The tree with flowers is to the left of the apple seeds.
2. The tree with only leaves is not before the tree with flowers.
3. The tree with flowers and tree with fruit are on an end.
4. The seeds are before the tree with leaves.

Problem 2

1. The tree with the fruit is not beside the seeds.
2. The flowers are either first or last.
3. There are two trees between the flowers and the seeds.
4. The tree with leaves is to the left of the seeds.

Problem 3

1. The tree that has flowers on it is on an end.
2. The picture with seeds is on the other end.
3. The tree with fruit is third.
4. The tree with just leaves is on the left of the tree with flowers.

Problem 4

1. The one with the long *e* sound made from two different vowels together is not first or third.
2. The one that rhymes with roots is not third or fourth.
3. The one with the double *e* has 3 trees after it.
4. The 2-syllable word is first or third.

Answers: 1. flowers, seeds, leaves, fruits; 2. flowers, fruits, leaves, seeds; 3. seeds, leaves, fruits, flowers; 4. seeds, fruits, flowers, leaves

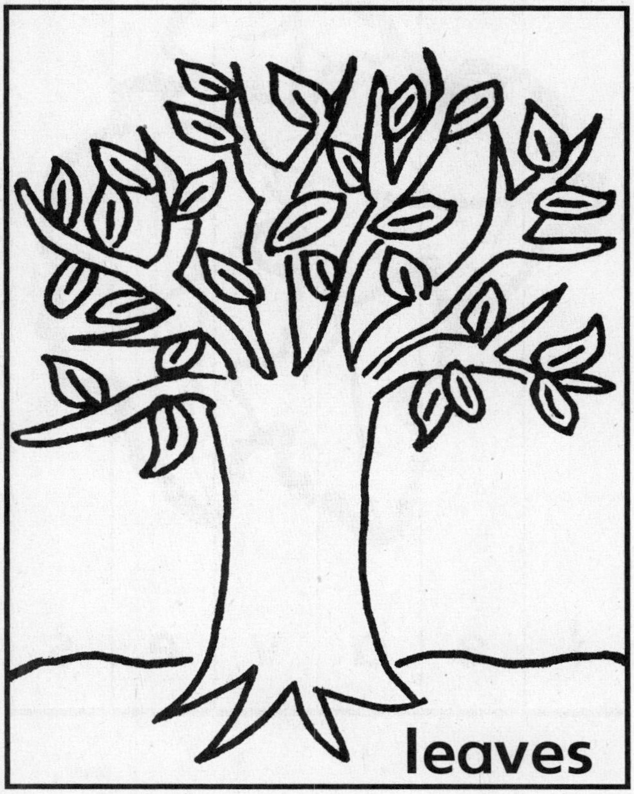
leaves

flowers

fruits

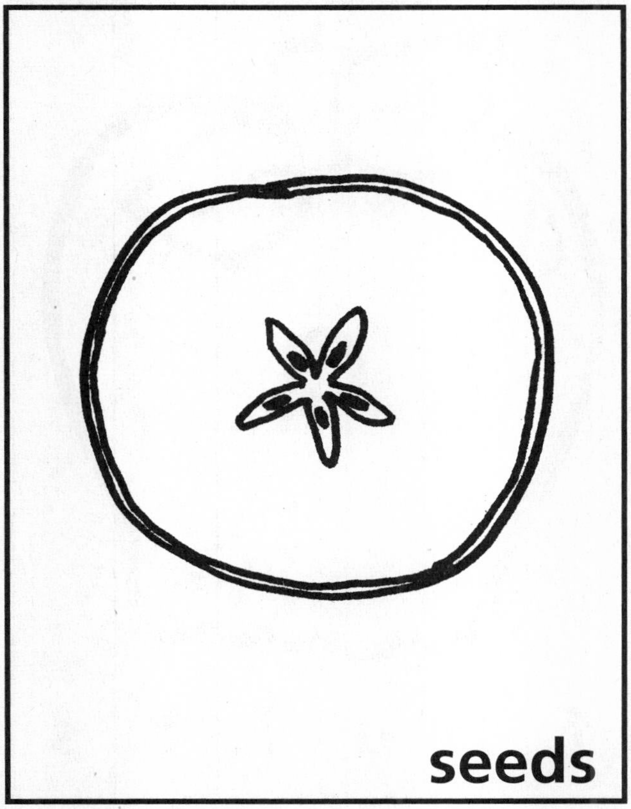
seeds

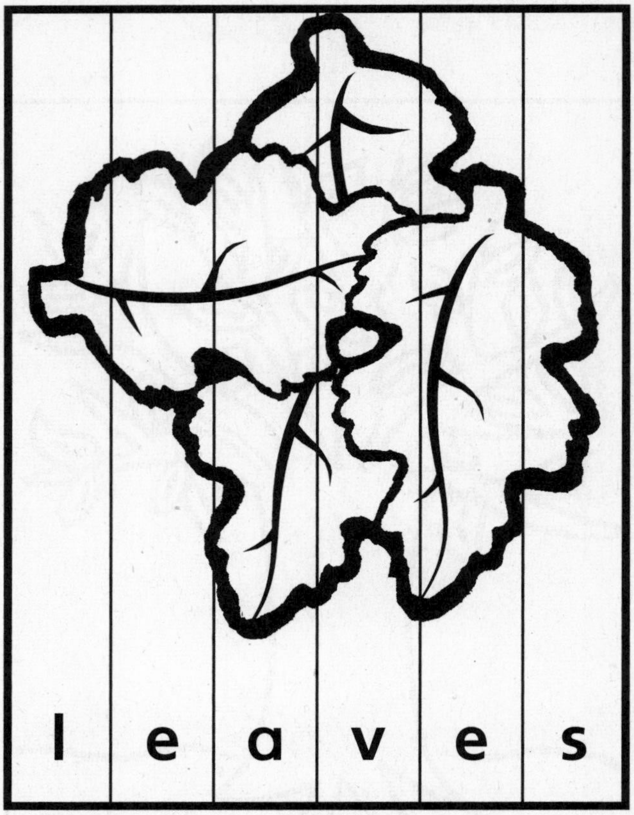

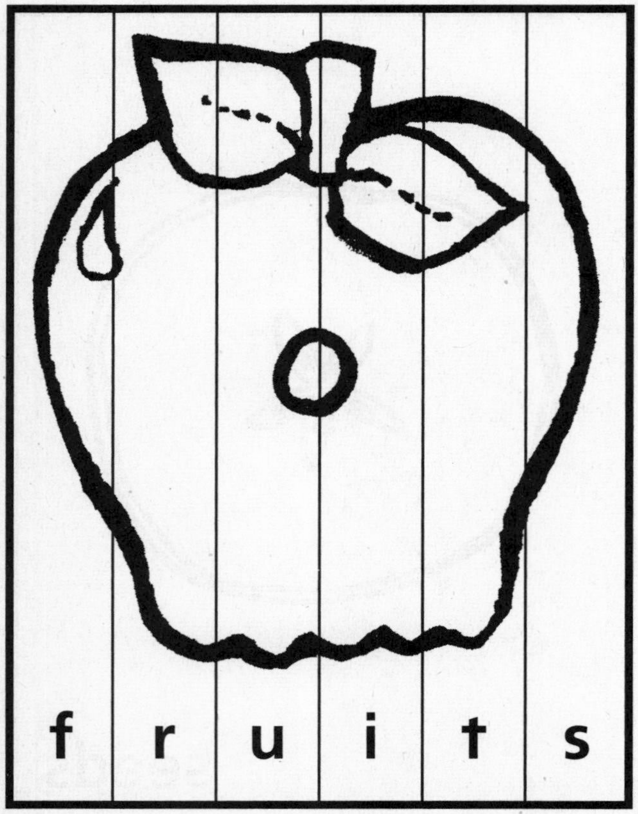

Use with p. 9
Plants Spin and Spell

i
u
r
f
s
t
o

v
a
e
l
s
e

w
o
l
f
r
e
o

Types of Animals

Animal Environments

Animal Needs

Animal Traits

Reptiles Are Animals

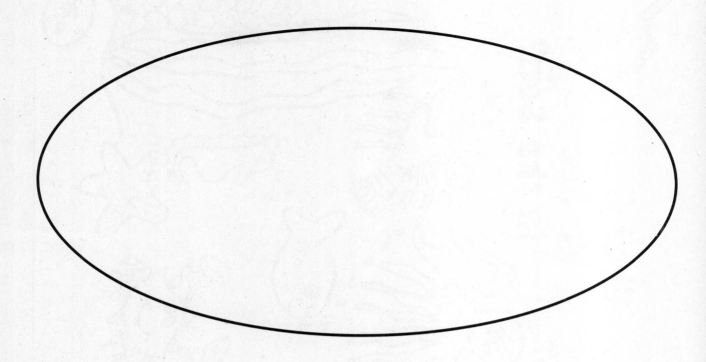

on ________________________________

under ________________________________

over ________________________________

through ________________________________

in ________________________, back home.

Sea Creatures

Red lobster

Blue whale

Yellow starfish

Green octopus

Purple fish

Brown sea horse

Orange crab

Black swordfish

White sand dollar

Gray stingray

What Can We See in the Sea?

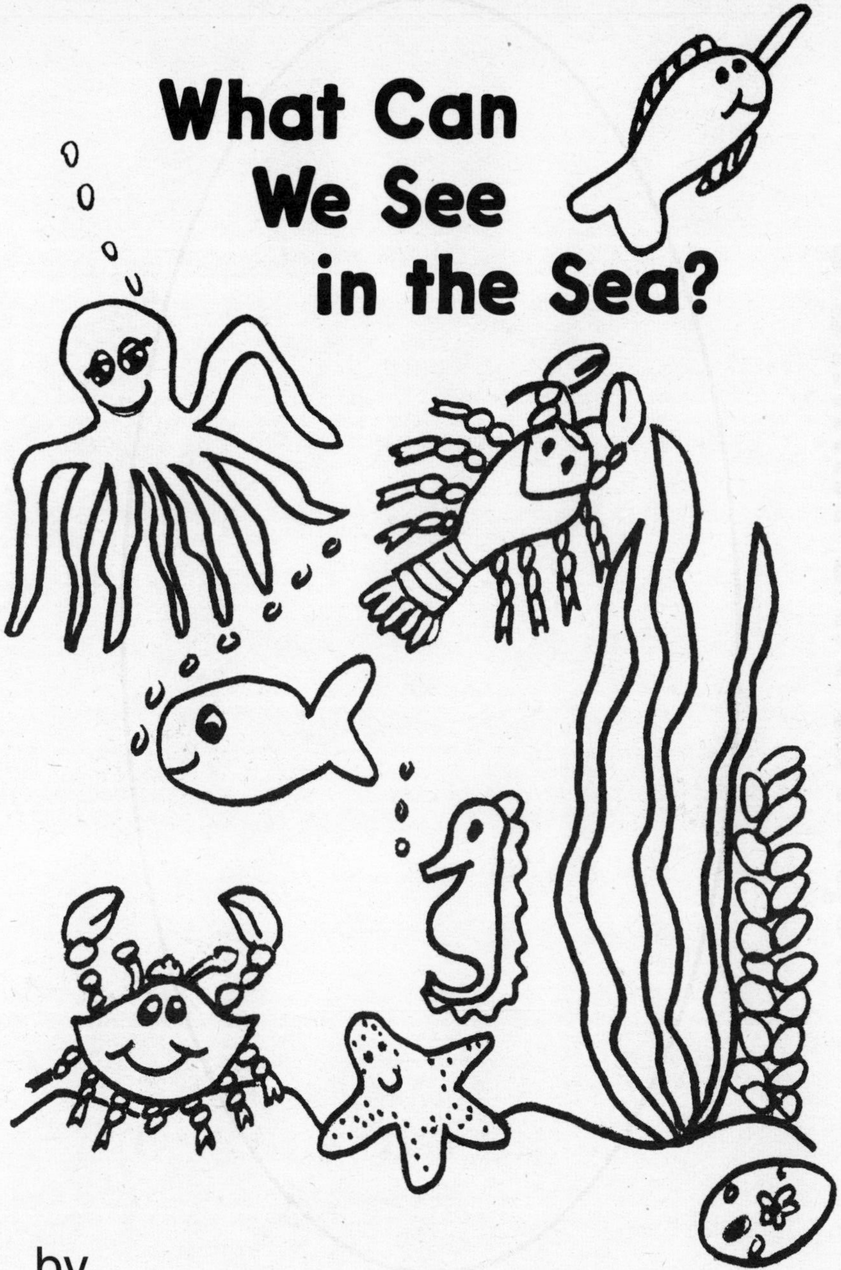

by ____________________

Use with p. 11
Ocean Creatures Are Animals

I see a ______________
looking at me.
It is ______________
______________.

I see a ______________
looking at me.
It is ______________
______________.

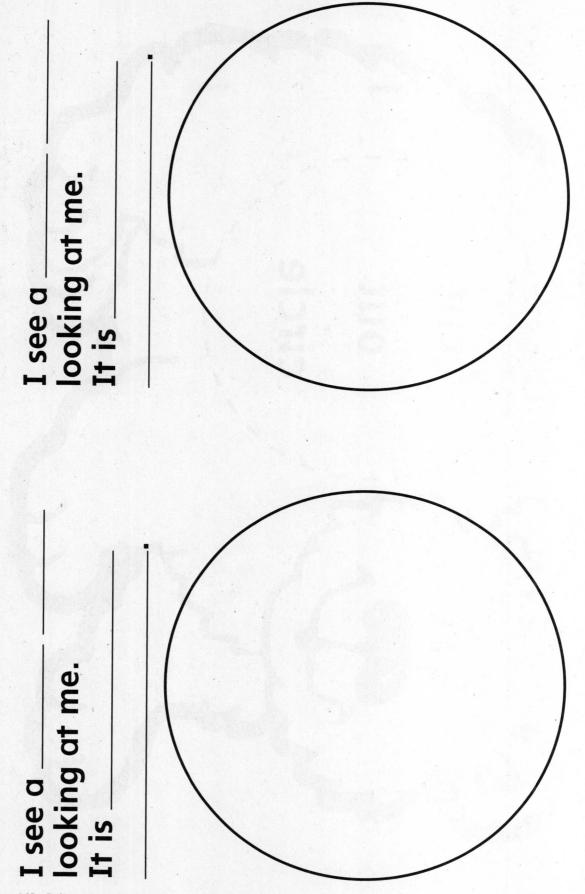

cut
out
circle

There Was an Old Bear

There was an old bear that swallowed some berries.
The berries weren't enough so he swallowed some nuts.

There was an old bear that swallowed some nuts.
The nuts weren't enough so he swallowed some bees.

There was an old bear that swallowed some bees.
The bees weren't enough so he swallowed some plants.

There was an old bear that swallowed some plants.
The plants weren't enough so he swallowed some fruit.

There was an old bear that swallowed some fruit.
The berries weren't enough so he swallowed a bunny.

There was an old bear that swallowed a bunny.
Now, as you can see, he is full in his tummy,
so he will sleep 'til it is sunny.

Animals Move

on land	in air	in water

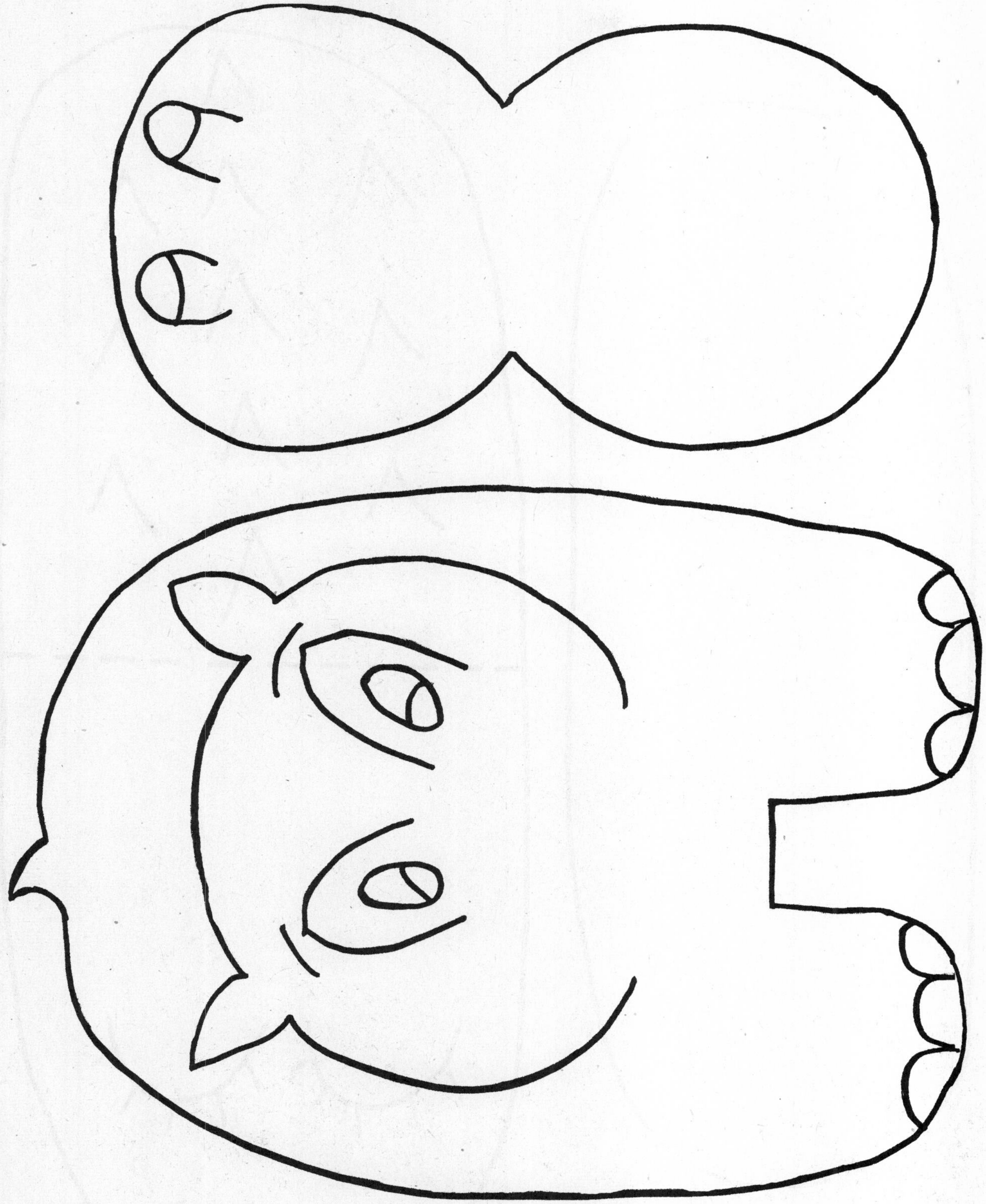

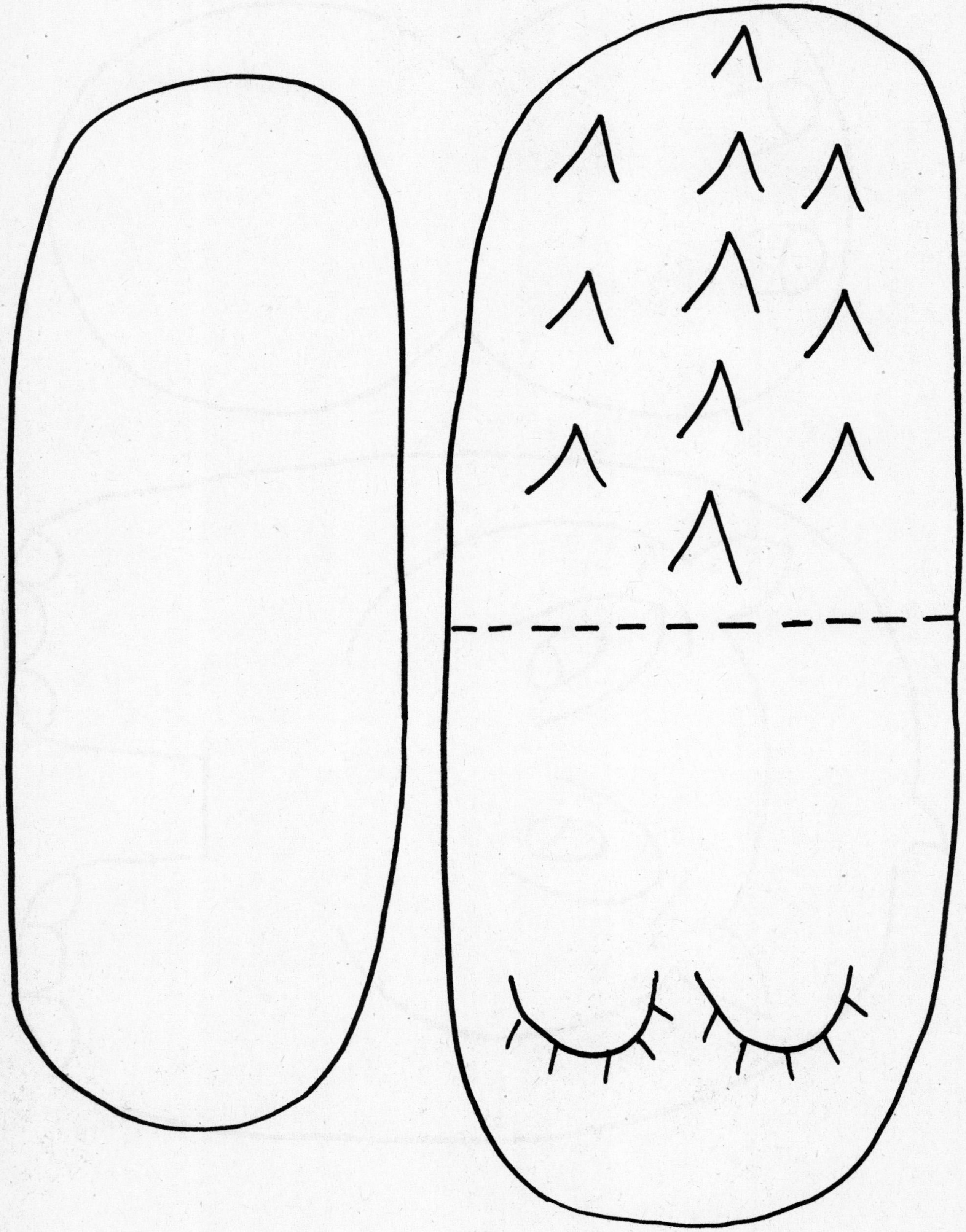

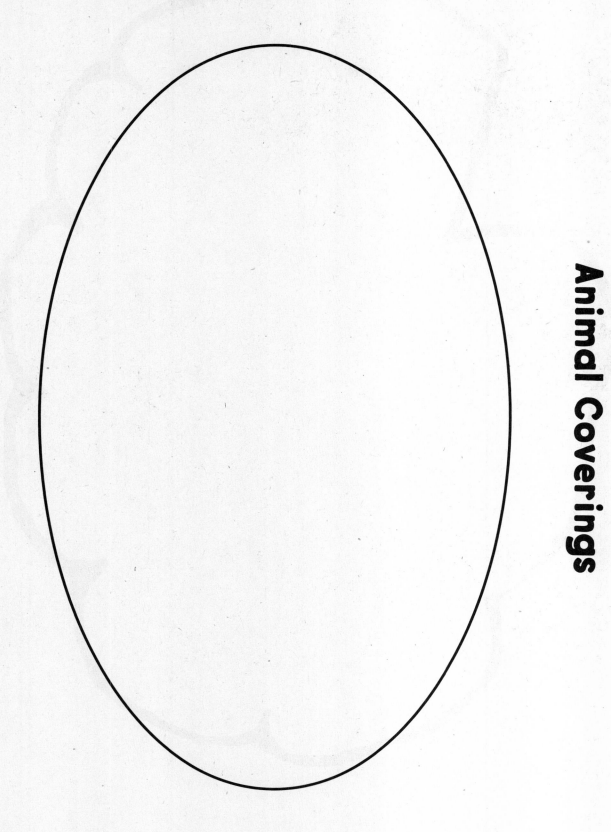

Animal Coverings

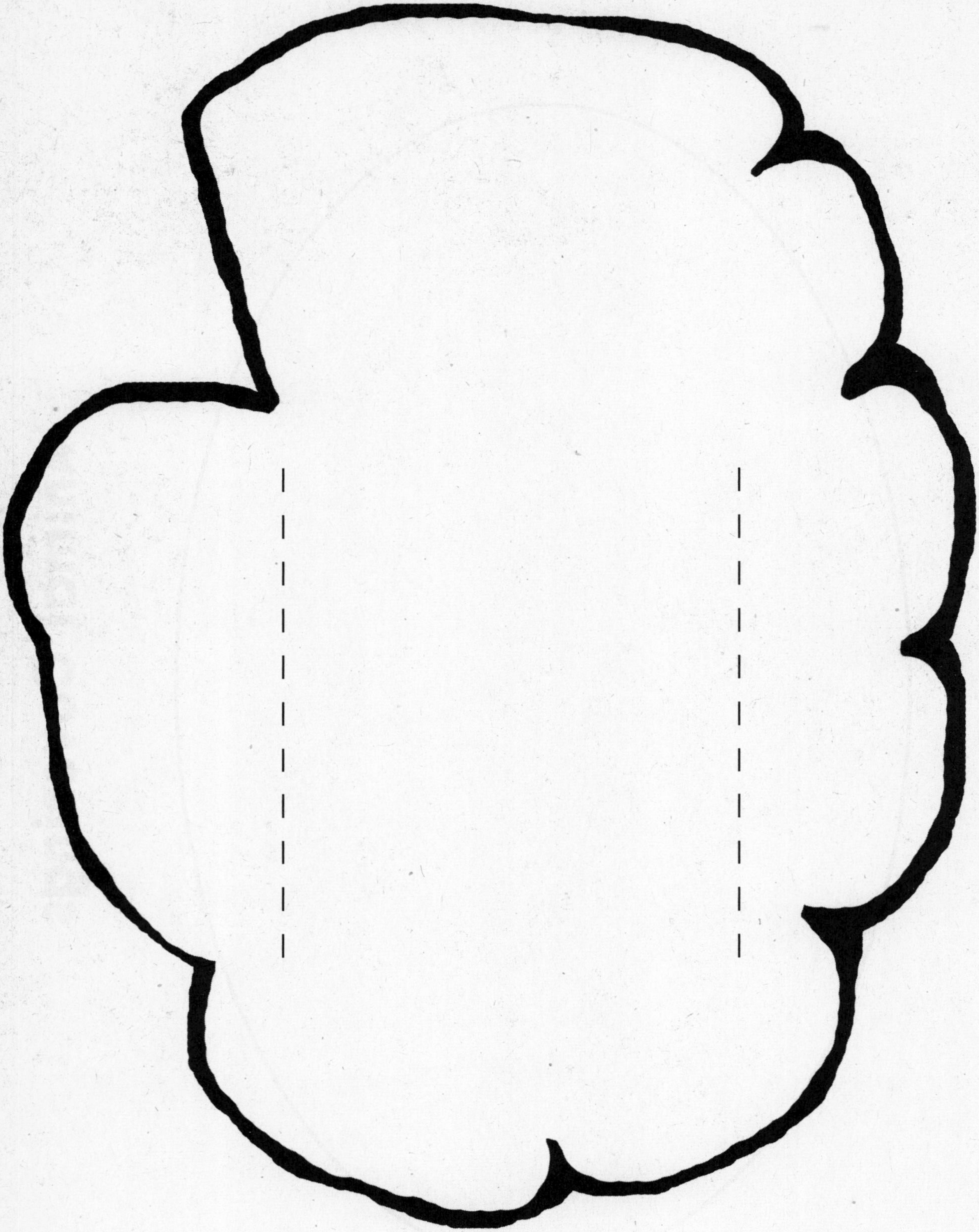

Life-Cycle Activity Strip

Plant and Animal Homes

Food Chains

Animal and Plant Needs

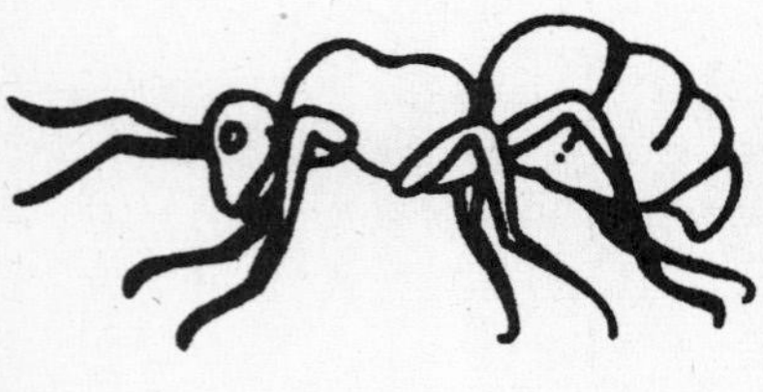

Insects

Traveling Through the Grasslands

in the grasslands
over the plains
under the trees
by the river

Forest Habitat

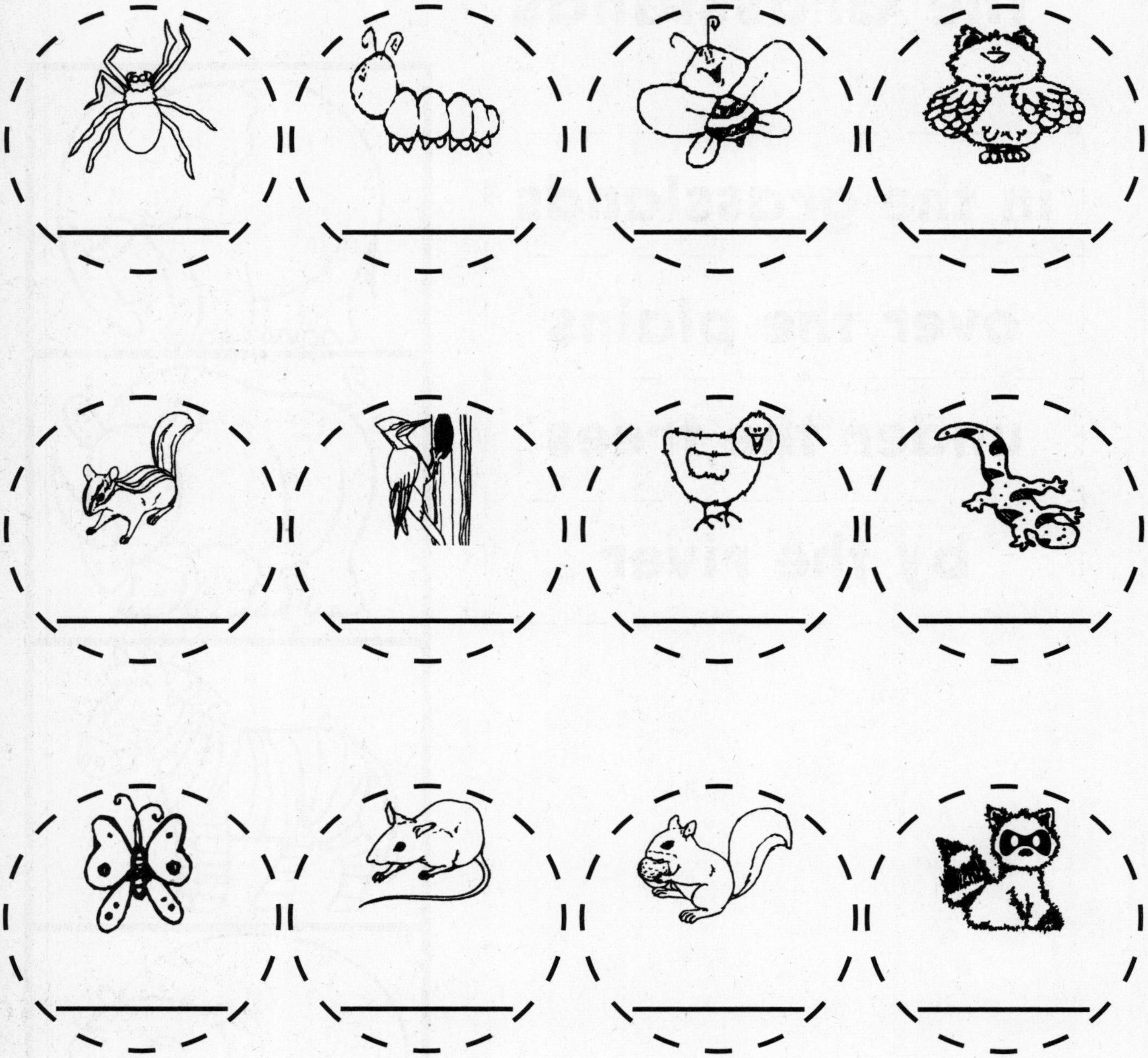

Use with p. 18
Forest Habitat

Rain-Forest Sentence Strip

Use these words in the top row: **grass, vines, plants, leaves**
Use these words in the bottom row: **ants, snakes, monkeys, butterflies**

Slithering Snake

Slithering
3 inches

Gigantic
$3\frac{1}{2}$ inches

Enormous
4 inches

Colossal
5 inches

Humongous
7 inches

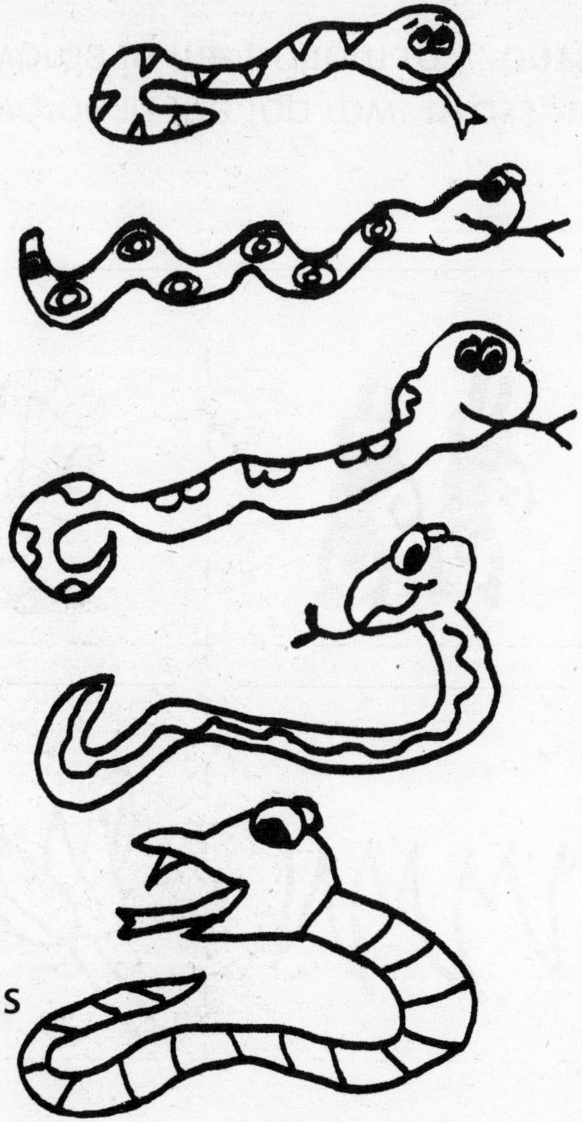

by ____________________

Slithering snake,
Slithering snake,
Oh, no, it's been swallowed by a . . .

1

Humongous snake,
Humongous snake,
B - U - R - P

Pardon Me!

6

____________________,

____________________,

______________ through the jungle.

5

____________________,

____________________,

______________ through the jungle.

____________________,

____________________,

Oh, no, it's been swallowed by a . . .

2

____________________,

____________________,

____________through the jungle.

____________________,

____________________,

____________through the jungle.

____________________,

____________________,

Oh, no, it's been swallowed by a . . .

3

____________________,

____________________,

Oh, no, it's been swallowed by a . . .

4

Lions Eat . . .

mice, lizards, turtles, fish, insects, birds, ostrich eggs, zebras, rhinoceroses, giraffes, young elephants, antelopes, and crocodiles.

On Sunday, he ate one

On Monday, he ate two

On Tuesday, he ate three

Paste under Wednesday

On Wednesday, he ate four

On Thursday, he ate five

On Friday, he ate six

On Saturday, he was full. He sat under a tree and slept from morning 'til night.

Use with p. 22
Plants and Animals Both Need. . .

Food-Chain Pyramid

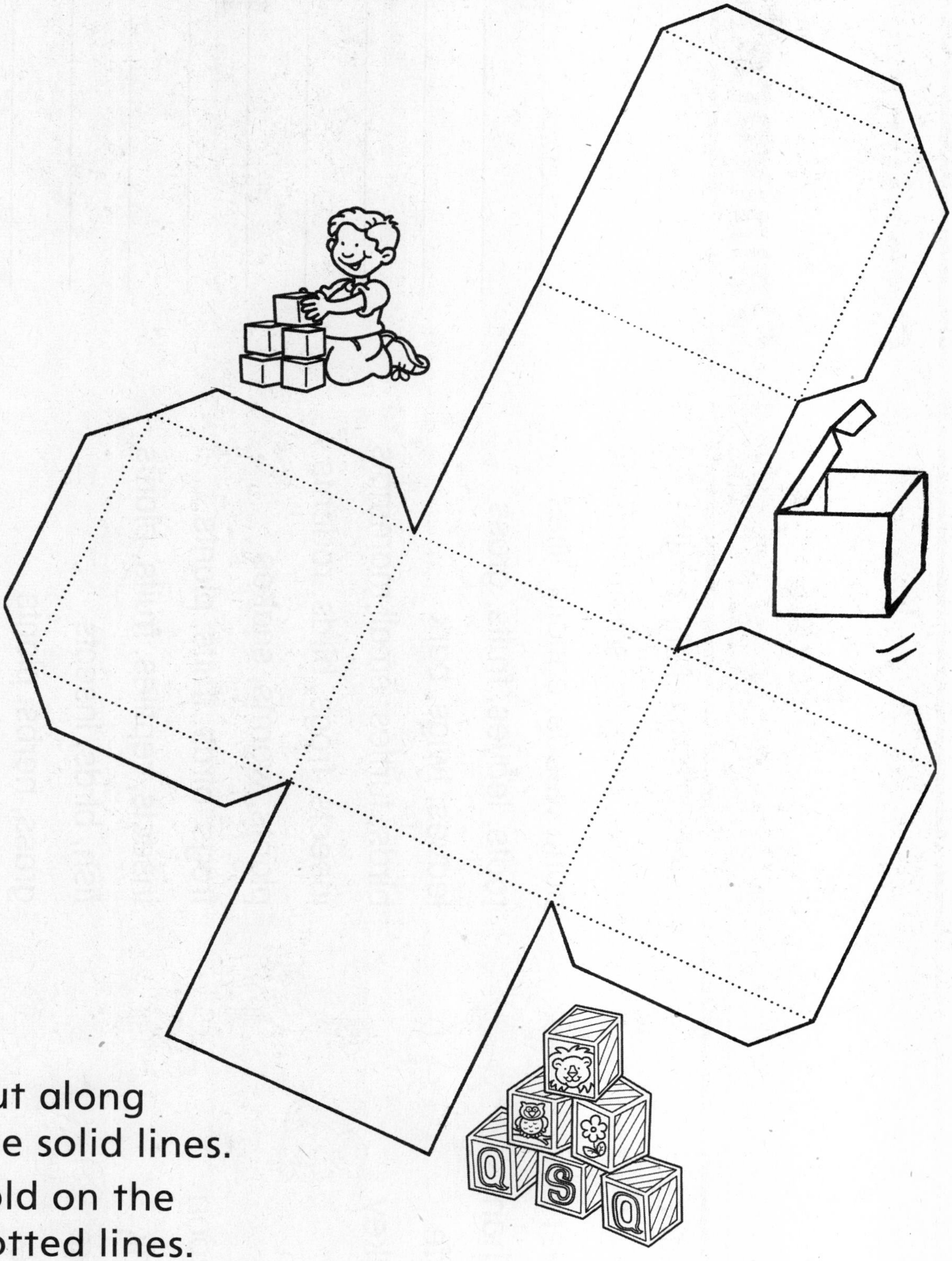

1. Cut along the solid lines.
2. Fold on the dotted lines.
3. Glue the tabs to create a cube.

Use with p. 23

Animal	Diet	Carnivore, Herbivore, or Omnivore?
alligator	insects, frogs, fish, snails	______
bear	plants, insects, small mammals	______
beaver	pond weeds, cattails, lilies	______
elephant	roots, leaves, fruits, grass	______
giraffe	leaves, twigs, bark	______
monkey	birds, turtles, small mammals	______
owl	insects, frogs, birds, rodents	______
pig	plants, worms, snakes	______
raccoon	frogs, birds, fruits, plants	______
skunk	insects, reptiles, fruits, plants	______
snake	fish, birds, insects	______
turtle	grass, herbs, plants	______

Use with p. 23

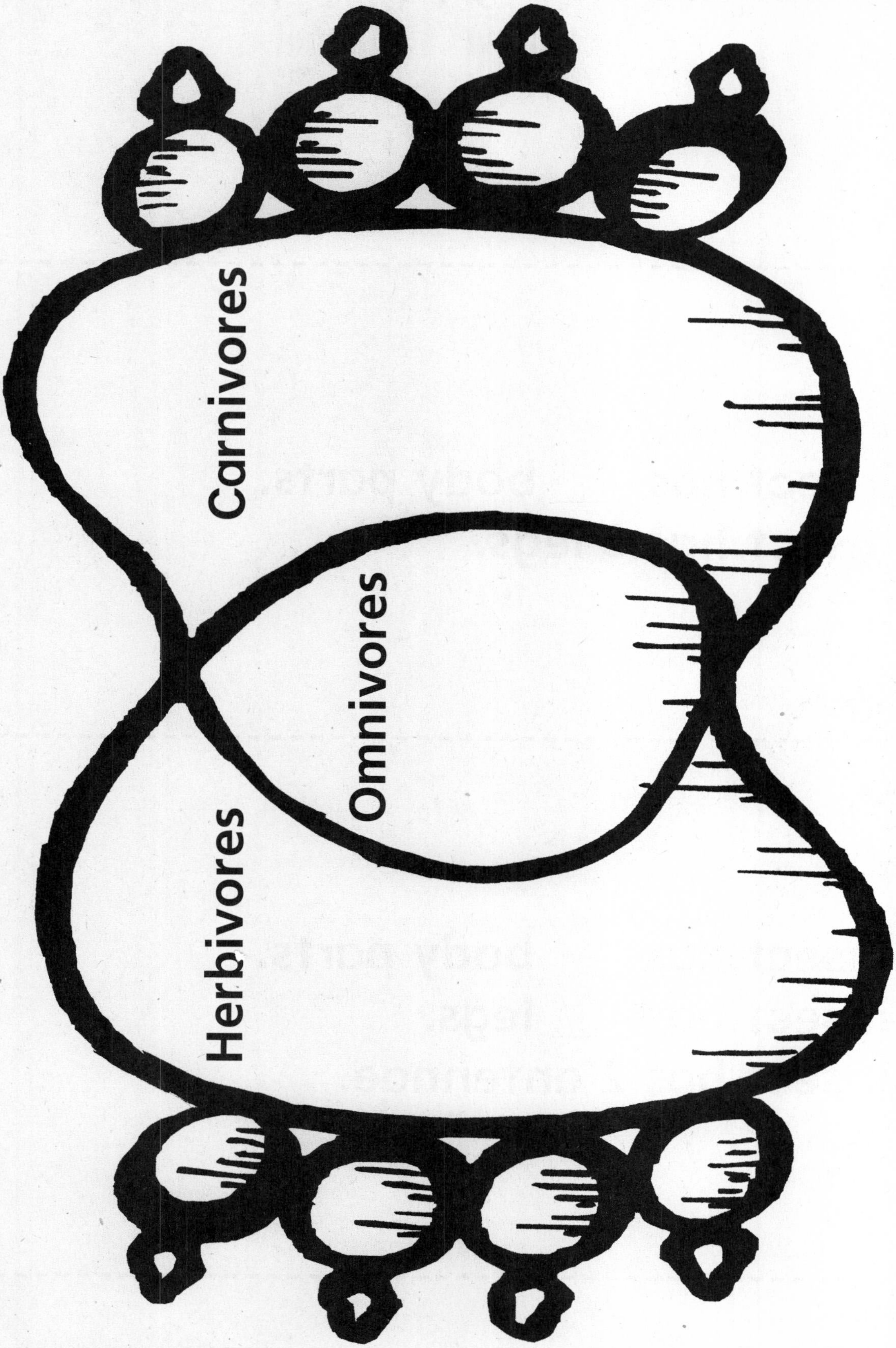
Carnivores
Omnivores
Herbivores

An insect has 3 body parts.

1

An insect has ___ body parts.
An insect has 6 legs.

2

An insect has ___ body parts.
An insect has ___ legs.
An insect has 2 antennae.

3

Use with p. 24
Design an Insect

An insect has ___ body parts.
An insect has ___ legs.
An insect has ___ antennae.
Most insects have 2 eyes.

4

An insect has ___ body parts.
An insect has ___ legs.
An insect has ___ antennae.
Most insects have ___ eyes.
Some insects have wings.

5

An insect has ___ body parts.
An insect has ___ legs.
An insect has 2 antennae.
Most insects have ___ eyes.
Some insects have wings.
This is my insect.

6

Spin an Insect

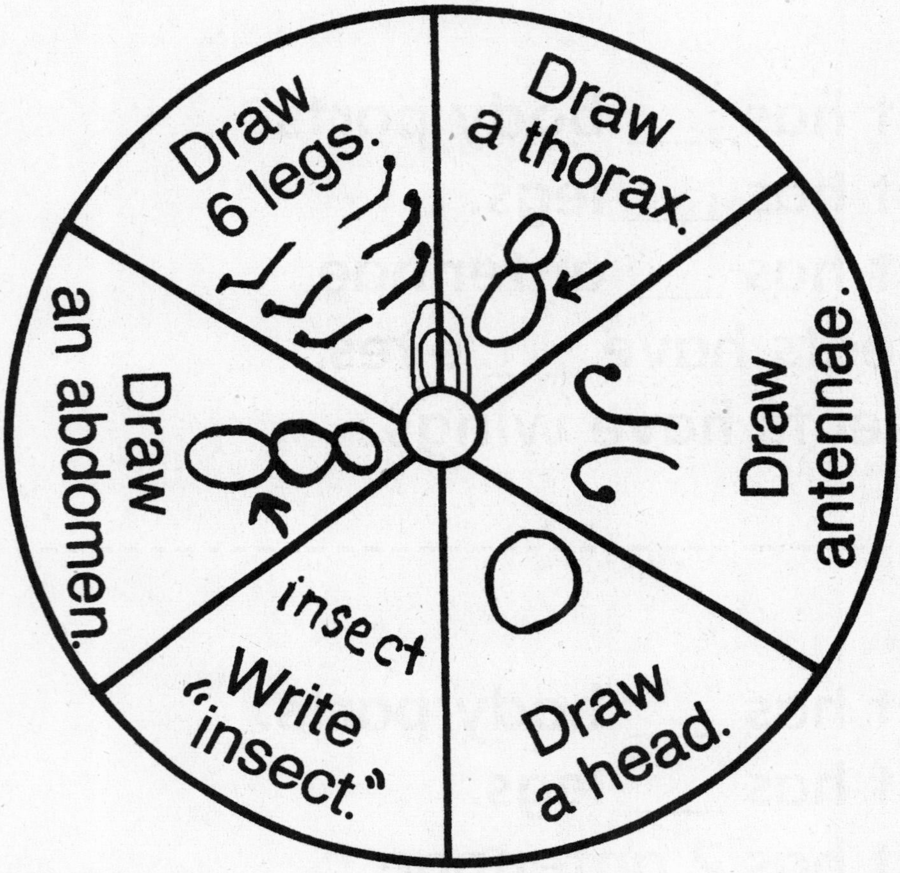

Tally Box: Make a tally mark each time a spin cannot be used.

Draw another insect and compare the number of tally marks.

Insect #I

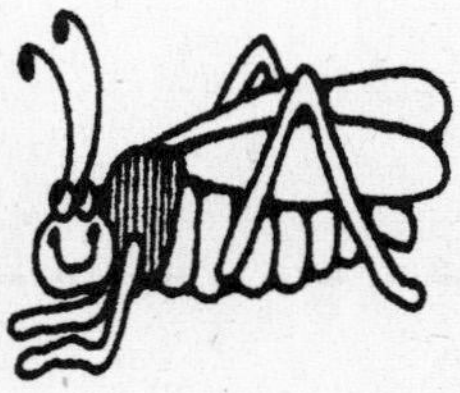

Insect #2

The Four Seasons

Changes in Seasons

Measuring Weather

Changes in Weather

____

is when

____

____

____

____

by ________________

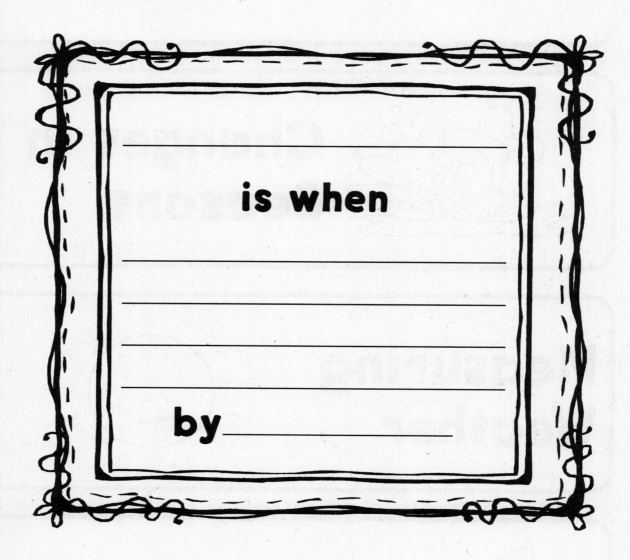

Use with **p. 26**
Seasons Quilt

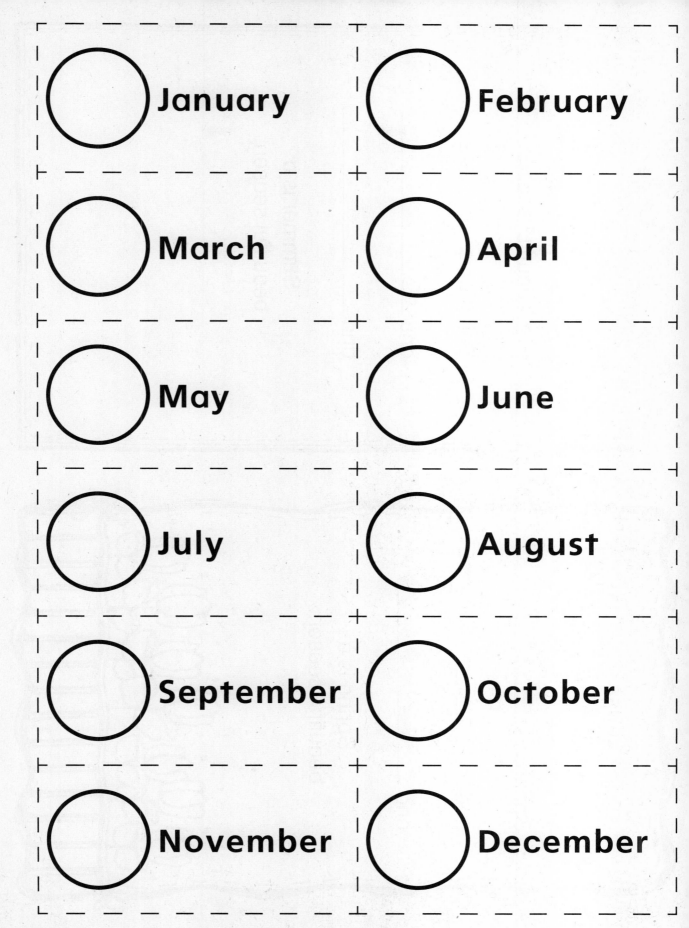

◯ January	◯ February
◯ March	◯ April
◯ May	◯ June
◯ July	◯ August
◯ September	◯ October
◯ November	◯ December

Spring is a
beautiful season.

Spring is ______________

and ____________________

and ____________________.

Spring is a
beautiful season.

Summer is a
beautiful season.

Summer is ______________

and ____________________

and ____________________.

Summer is a
beautiful season.

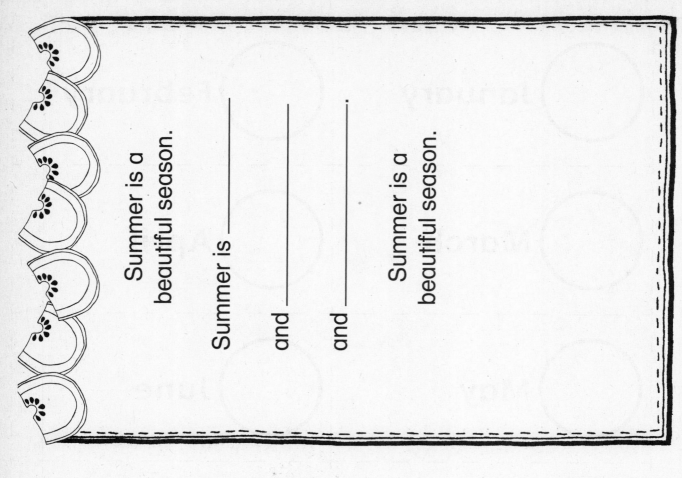

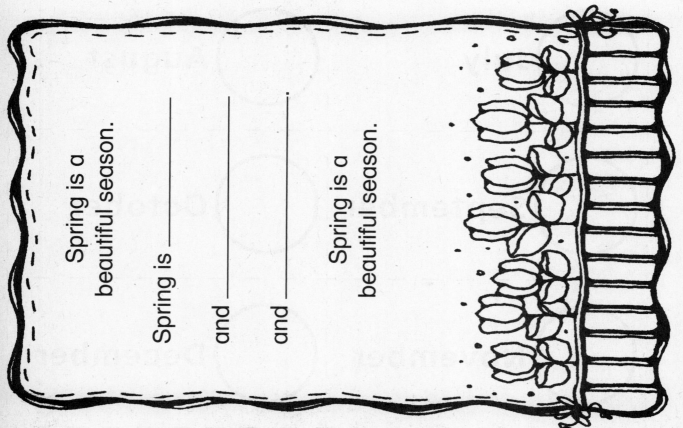

Fall is a
beautiful season.

Fall is ______

and ______

and ______.

Fall is a
beautiful season.

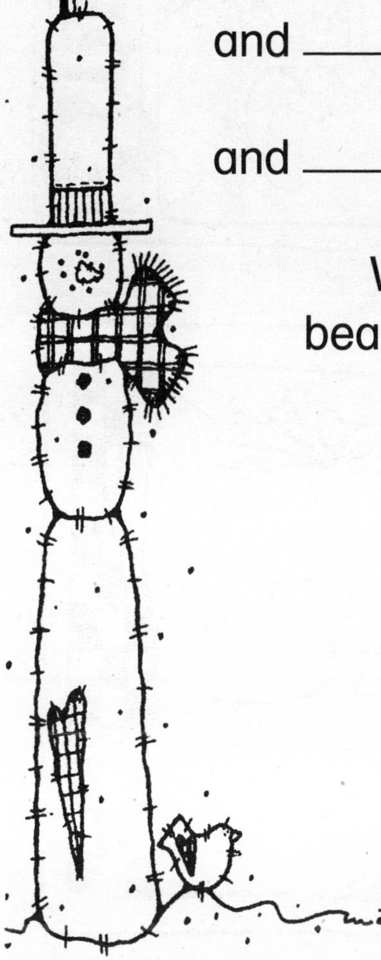

Winter is a
beautiful season.

Winter is ______

and ______

and ______.

Winter is a
beautiful season.

Use with p. 27
Seasons Logic Line-Up

Problem 1

Put the seasons in order, beginning with winter.
Put the seasons in order, beginning with summer.
Put the seasons in order, beginning with spring.
Put the seasons in order, beginning with autumn.

Problem 2

The two seasons that begin with the same letter are first and last.
The season that is also called fall is second.
The two seasons that end with "r" are next to each other.

(spring, autumn, winter, summer)

Problem 3

Put the seasons in alphabetical order.

(autumn, spring, summer, winter)

Problem 4

Put the seasons in order from hot to cold.
Put the seasons in order from cold to hot.

(summer, spring, autumn, winter; winter, autumn, spring, summer)

Problem 5

The season with Thanksgiving is second.
The season with Valentine's Day is last.
The season in which school vacation begins is third.
The season with the 4th of July is first.

(summer, autumn, spring, winter)

Seasons and Our Clothing

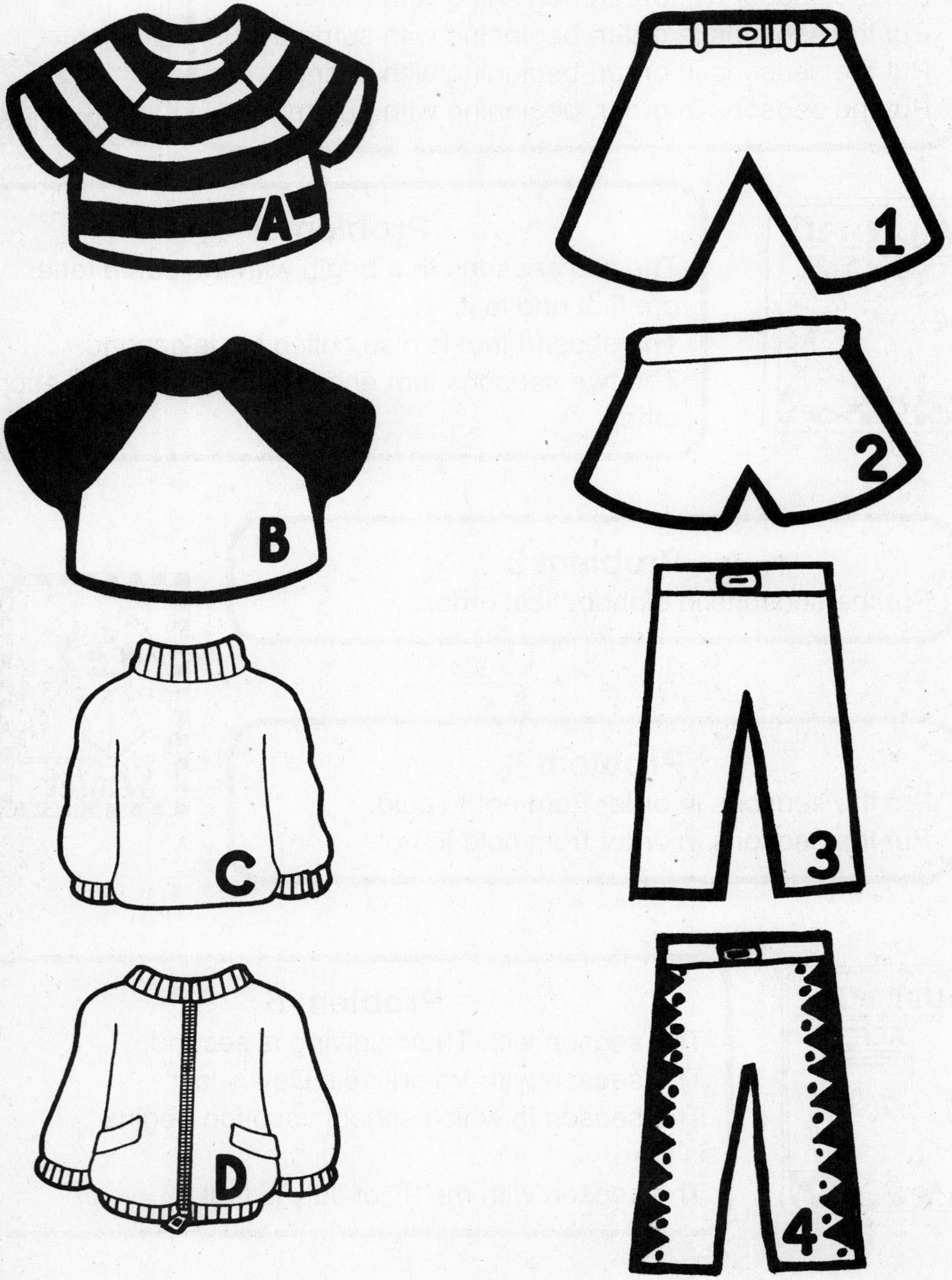

In the ________________,

the trees ____________

____

________________________.

Use with p. 28

Hibernating Animals

(Tune: "Alouette")

Hibernation, time for hibernation,
Hibernation, time to go to sleep.
In the winter, where's the raccoon?
Sleeping in a big old tree.
Where's the raccoon? In a tree.
O-o-o-o-o-oh
Hibernation, time for hibernation,
Hibernation, time to go to sleep.

Hibernation, time for hibernation,
Hibernation, time to go to sleep.
In the winter, where's the snail?
Sleeping in a small hard shell.
Where's the snail? In a shell.
O-o-o-o-o-oh
Hibernation, time for hibernation,
Hibernation, time to go to sleep.

Hibernation, time for hibernation,
Hibernation, time to go to sleep.
In the winter, where's the snake?
Sleeping under a big hard rock.
Where's the snake? Under a rock.
O-o-o-o-o-oh
Hibernation, time for hibernation,
Hibernation, time to go to sleep.

Use with p. 29
Hibernation

Hibernation, time for hibernation,
Hibernation, time to go to sleep.
In the winter, where's the bat?
Sleeping in a dark, cold cave.
Where's the bat? In a cave.
O-o-o-o-o-oh
Hibernation, time for hibernation,
Hibernation, time to go to sleep.

Hibernation, time for hibernation,
Hibernation, time to go to sleep.
In the winter, where's the turtle?
Sleeping down in the ground.
Where's the turtle? In the ground.
O-o-o-o-o-oh
Hibernation, time for hibernation,
Hibernation, time to go to sleep.

Hibernation, time for hibernation,
Hibernation, time to go to sleep.
In the winter, where's the frog?
Sleeping under a pile of mud.
Where's the frog? In the mud.
O-o-o-o-o-oh
Hibernation, time for hibernation,
Hibernation, time to go to sleep.

Winter Homes for Animals

A ____________

A ____________

A ____________

A ____________

A ____________

A ____________

Use with p. 29
Hibernation

in a ______________.	in a ______________.
under a ______________.	in a ______________.
in the ______________.	in the ______________.

Use with p. 29

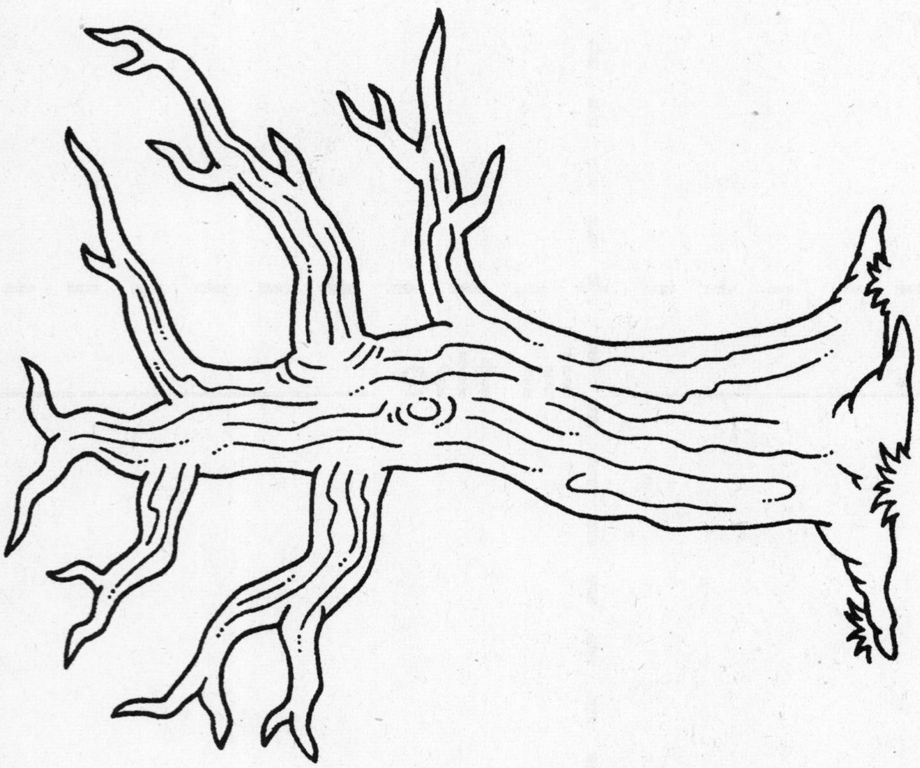

Measuring With Thermometers

Directions: Look at the thermometers next to the pictures. Set your thermometer to each temperature. Write the name of the season and the temperature under each picture. Use the picture to tell whether the weather is **snowy, windy, rainy,** or **sunny**.

winter: 30°F

fall: 50°F

spring: 70°F

summer: 90°F

Use with p. 30

°F
130
120
110
100
90
80
70
60
50
40
30
20
10
0
−10
−20
−30

I can't	I can't	I can't	I can't

because

because

because

because

Use with p. 32
Weather and Activities

Natural Resources

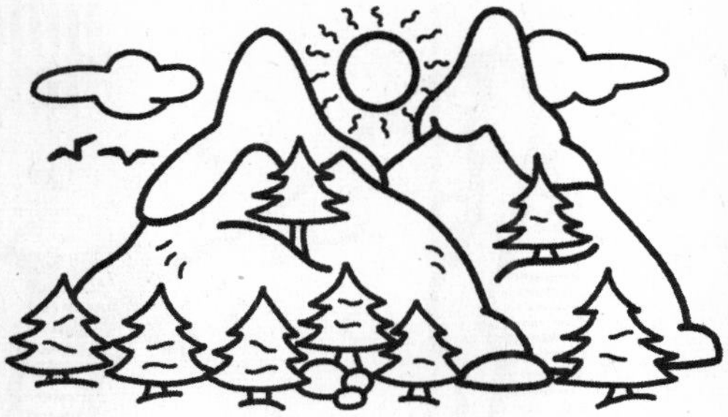

Using Natural Resources

Let's Study Fossils

Dinosaurs Are Great

Rocks Are Natural Resources

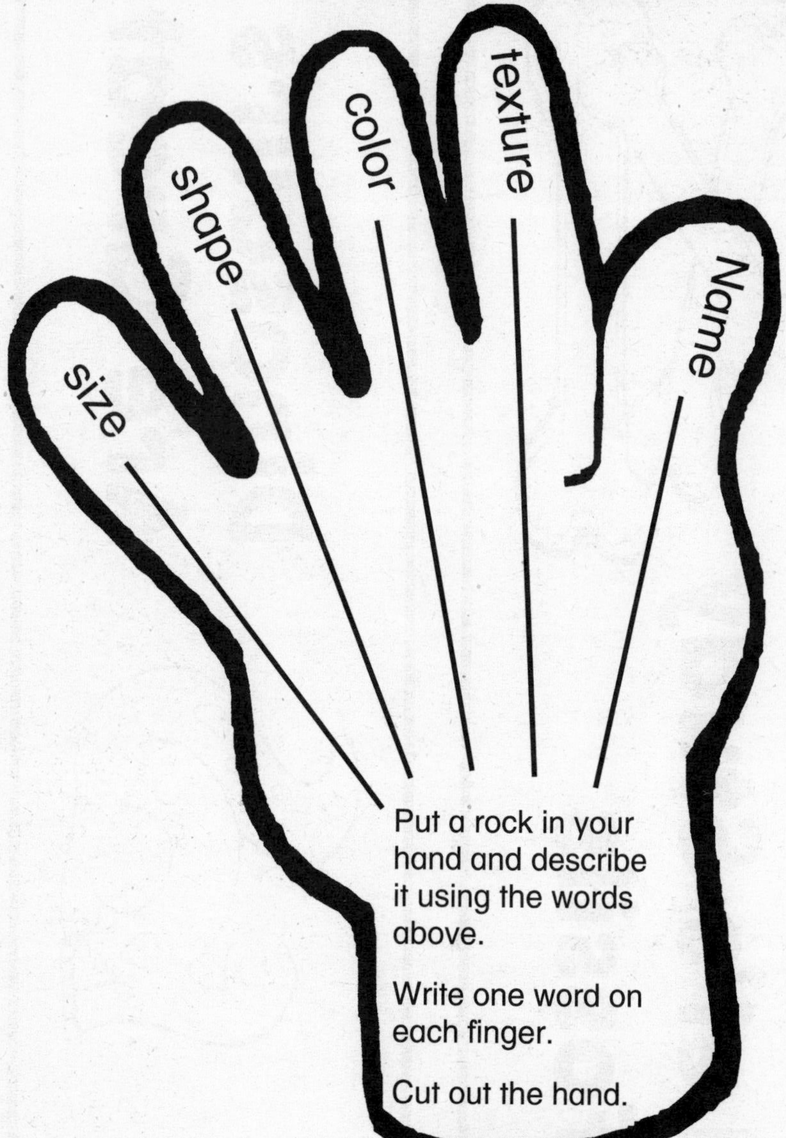
size
shape
color
texture
Name
Put a rock in your hand and describe it using the words above.
Write one word on each finger.
Cut out the hand.

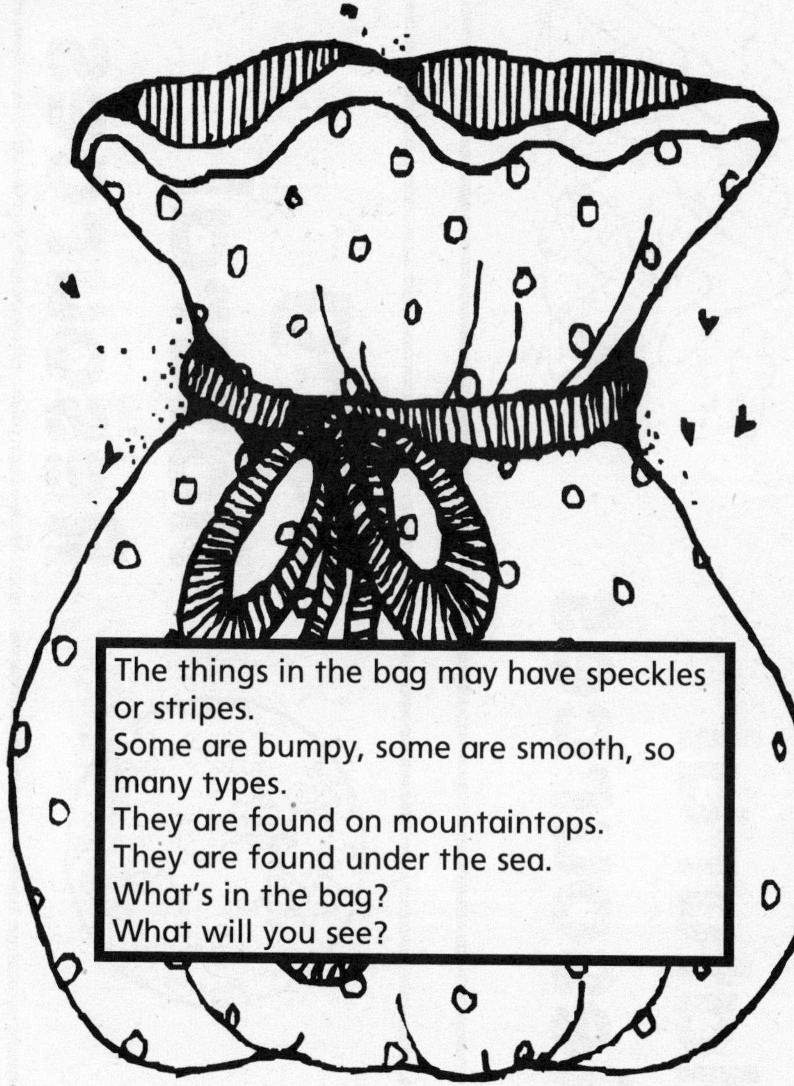
The things in the bag may have speckles or stripes.
Some are bumpy, some are smooth, so many types.
They are found on mountaintops.
They are found under the sea.
What's in the bag?
What will you see?

Water Cycle

Put each word in the correct box:

precipitation
evaporation
condensation

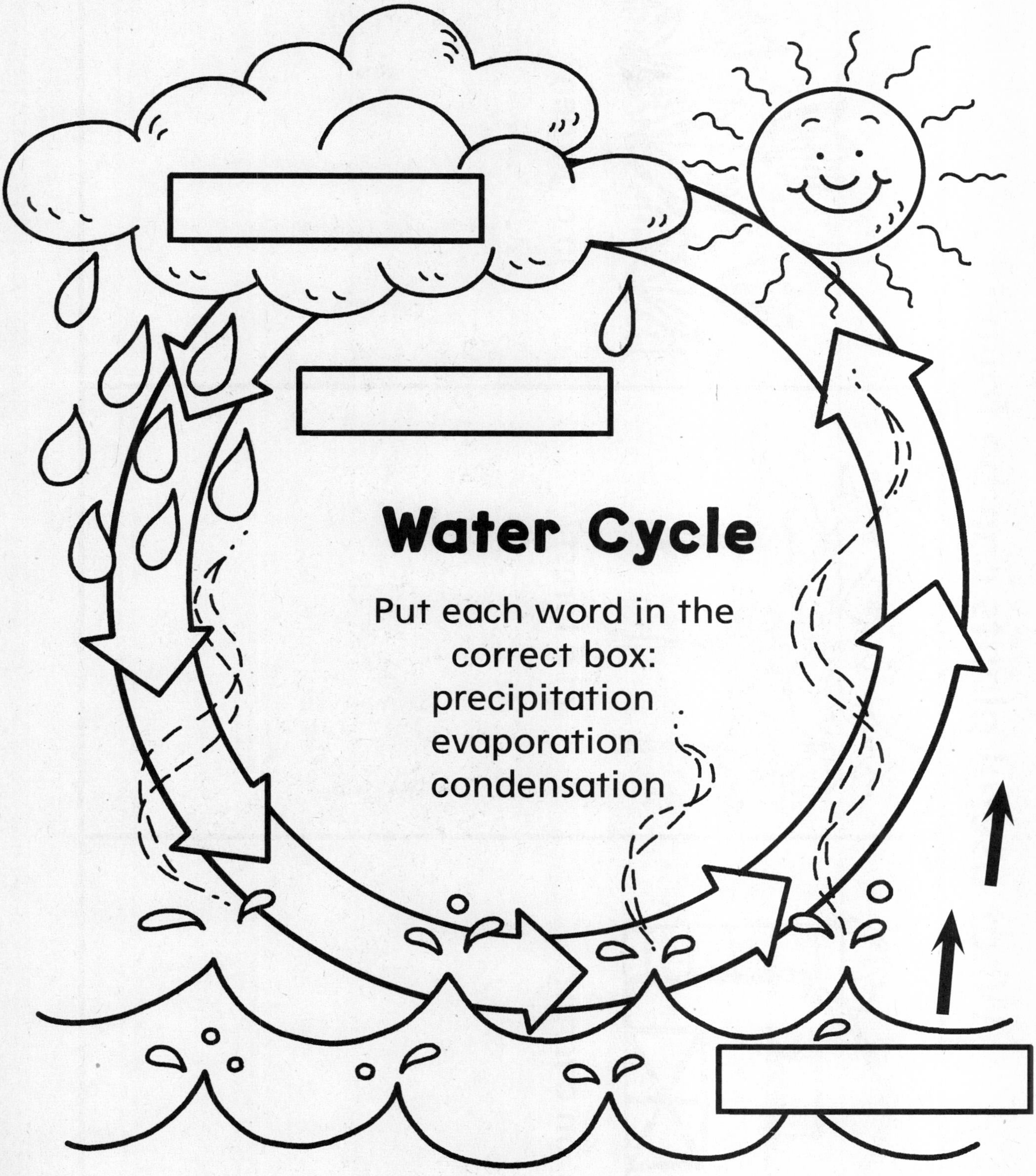

What useful plants can be found . . .

on a farm?

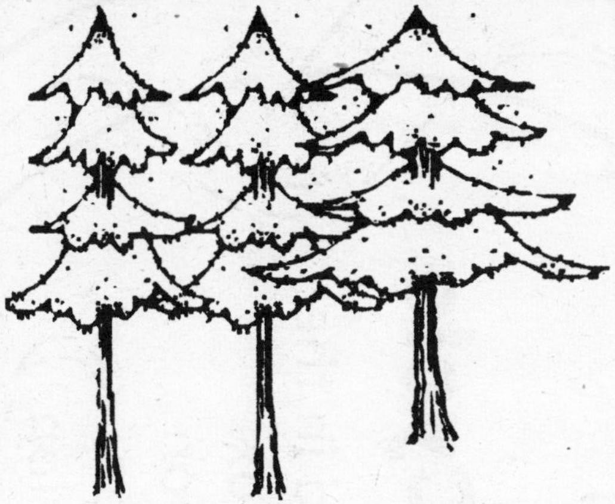

in a forest?

in a garden?

You grow ____________

You grow ____________

You grow ____________

You grow ____________

to make ____________

to make ____________

to make ____________

to make ____________

Use with p. 36
The Importance of Plants

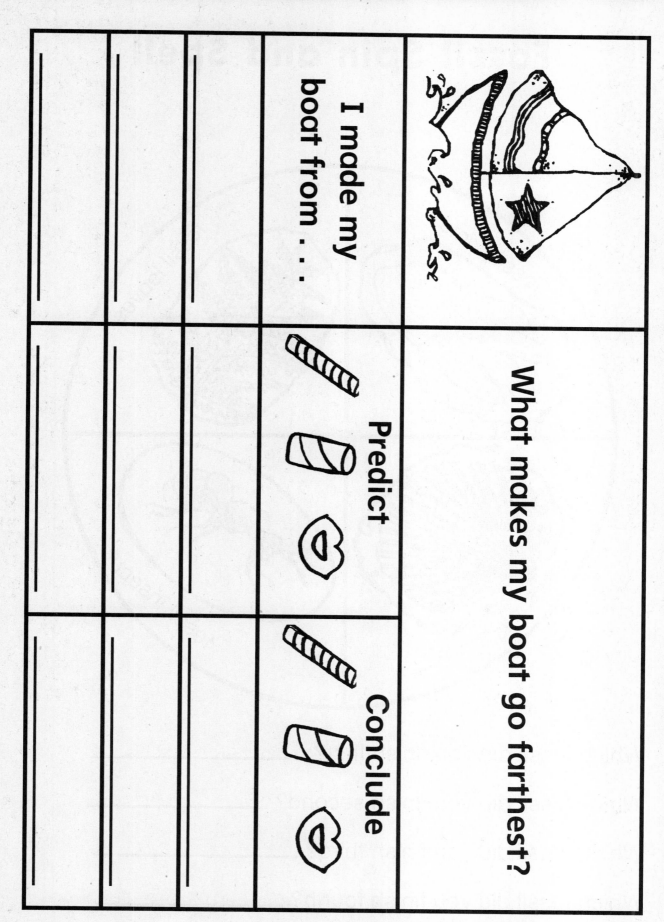

What makes my boat go farthest?

I made my boat from . . .	Predict 	Conclude 
________	________	________
________	________	________
________	________	________

Fossil Spin and Spell

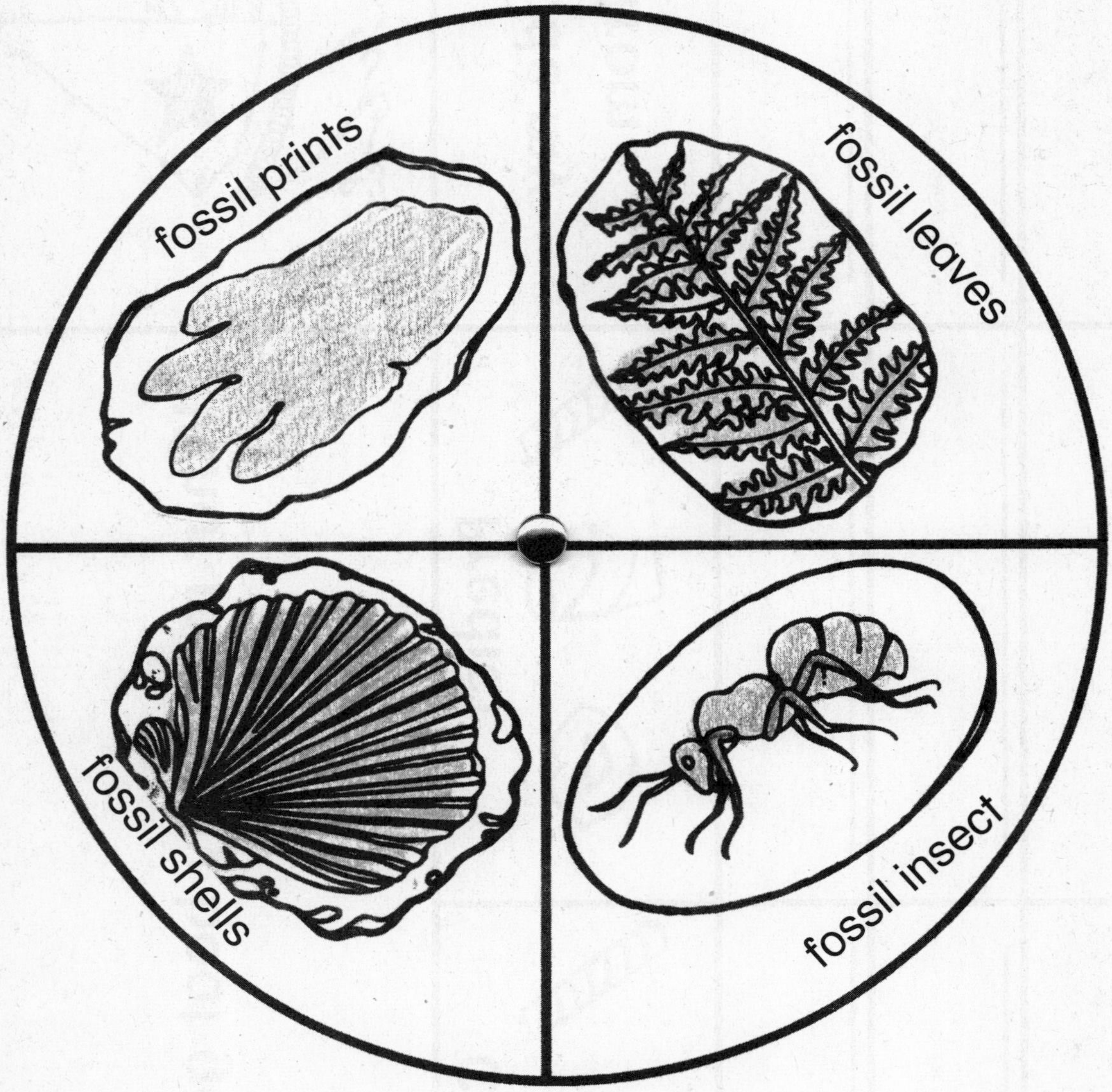

Which fossil did you finish first? ____________________

Which fossil did you finish second? ____________________

Which fossil did you finish third? ____________________

Which fossil did you finish fourth? ____________________

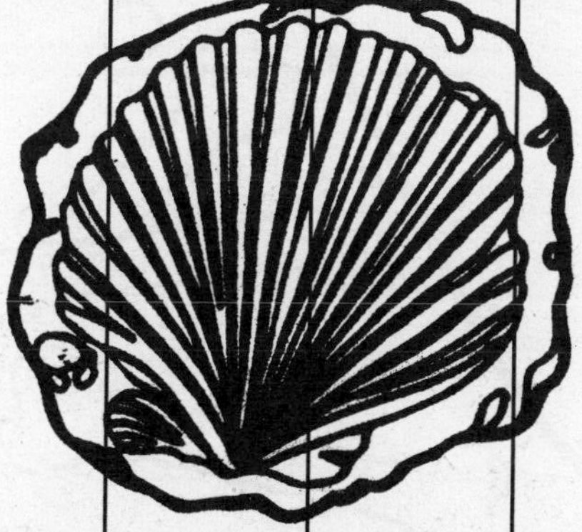

s h e l l s

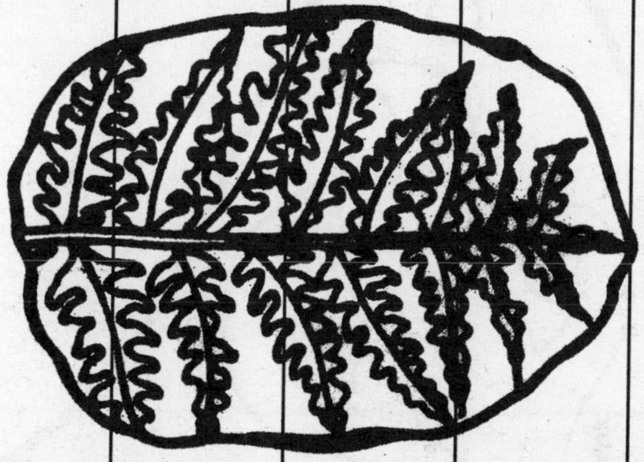

l e a v e s

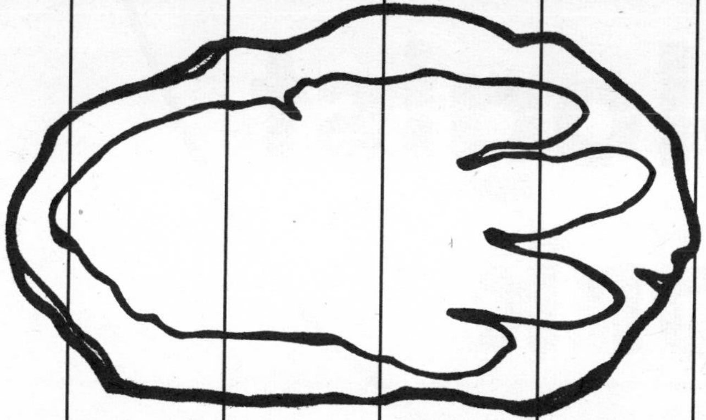

p r i n t s

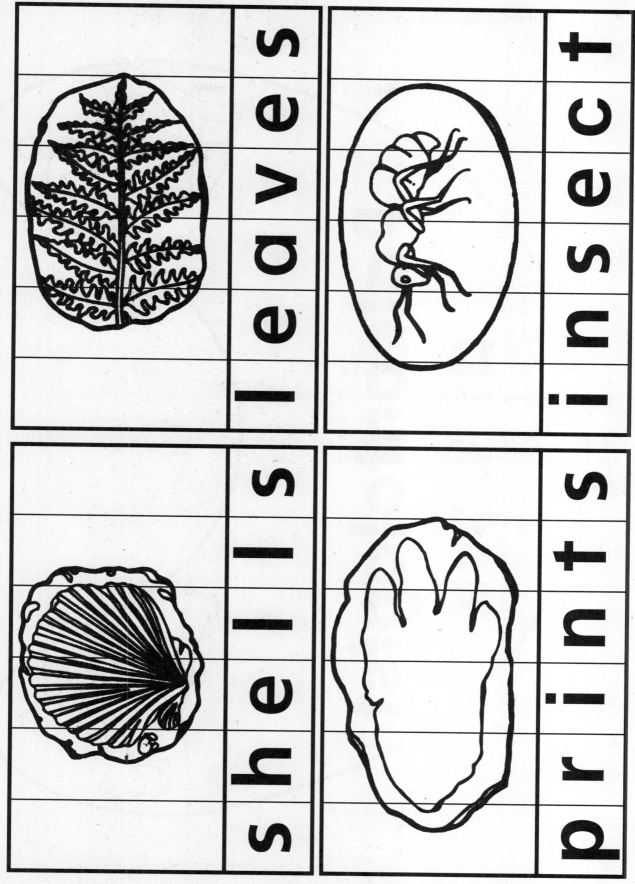

i n s e c t

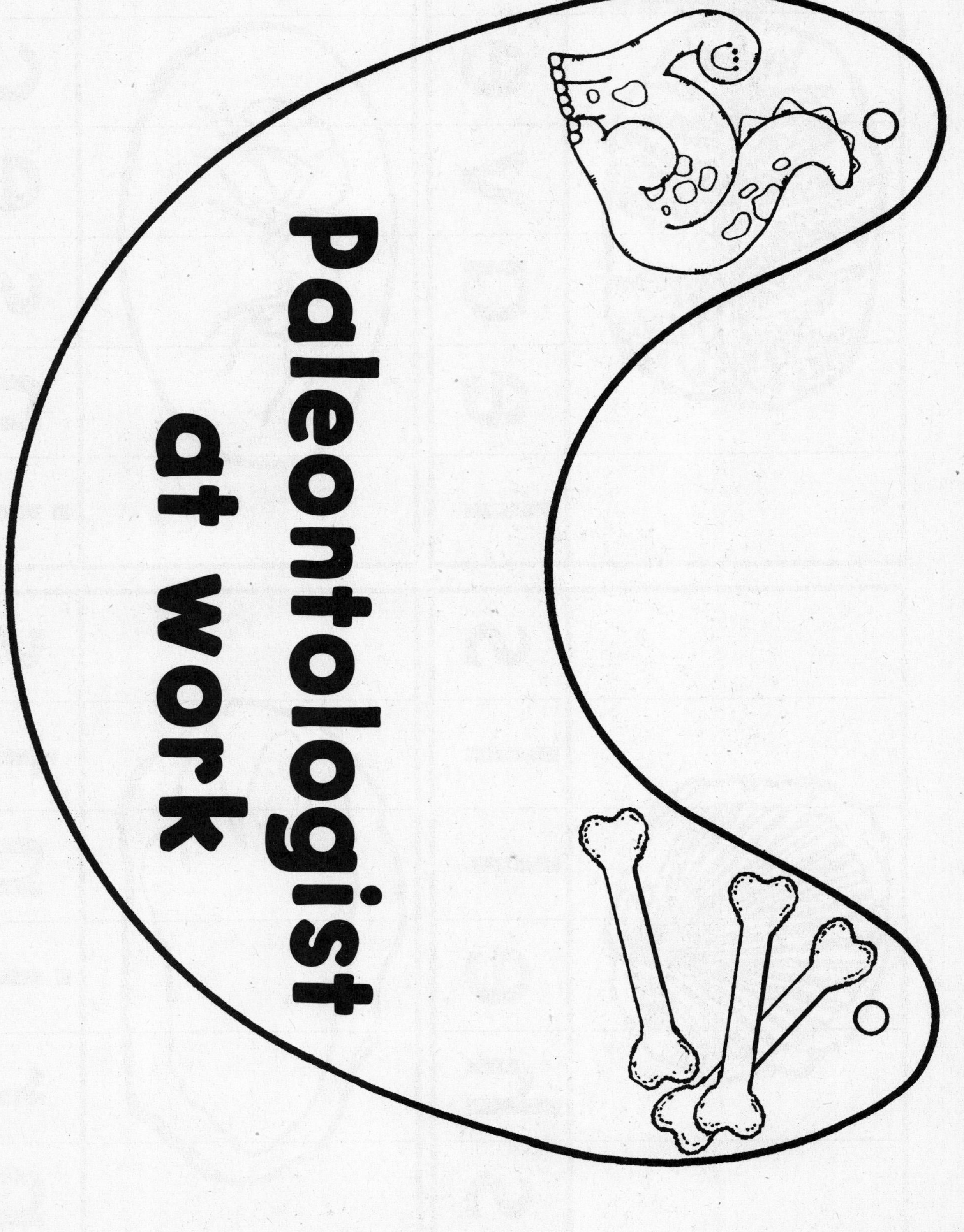
Paleontologist
at work

Dinosaur Footprints

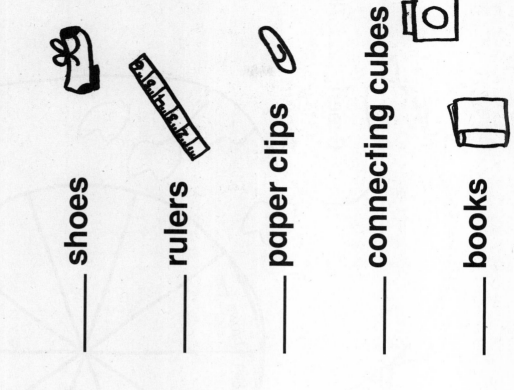

_____ shoes 

_____ rulers 

_____ paper clips 

_____ connecting cubes 

_____ books 

Use with p. 40

Dinosaur Probability

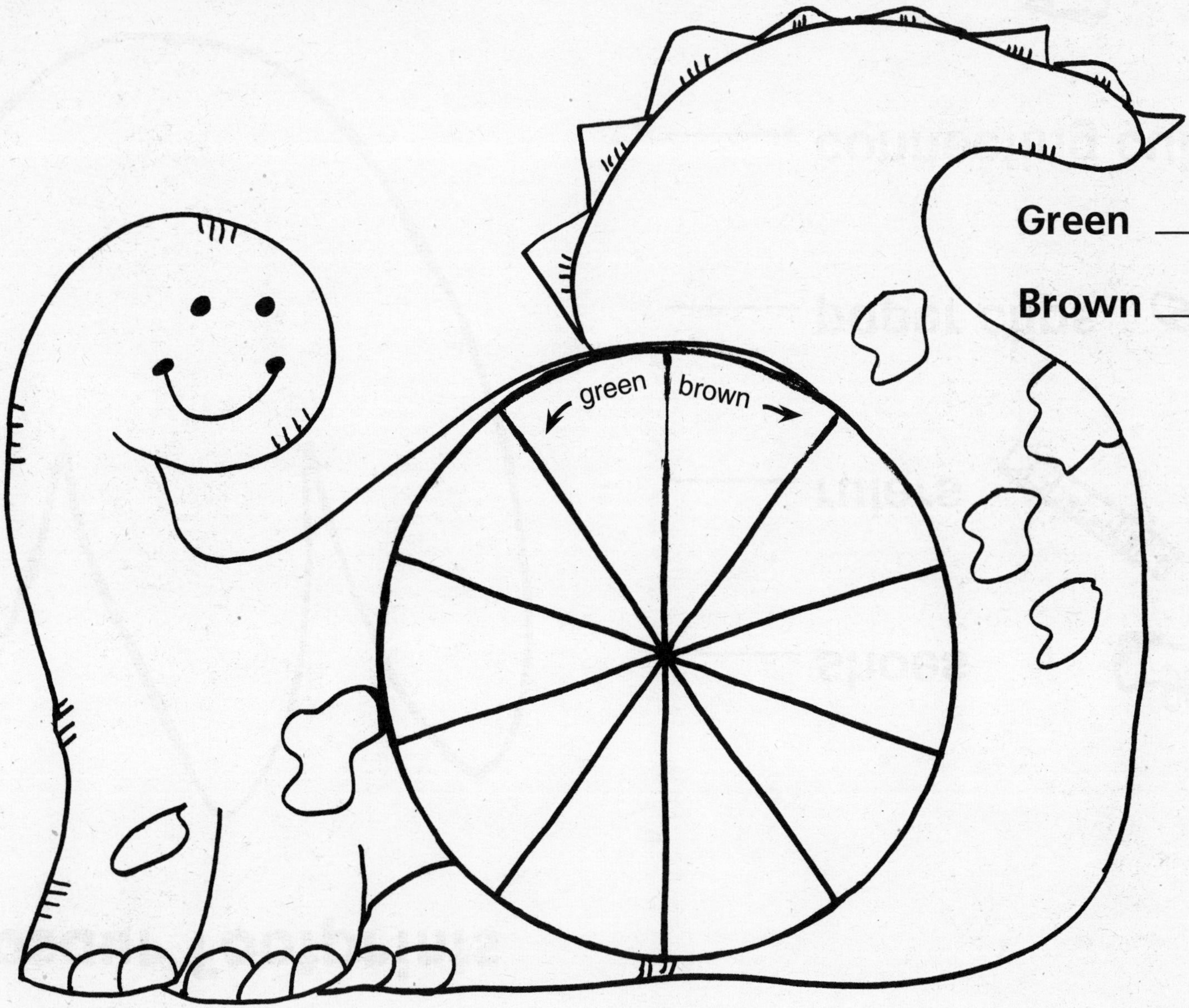

Green _____/10

Brown _____/10

Lambeosaurus

Ankylosaurus

Earth's Movements

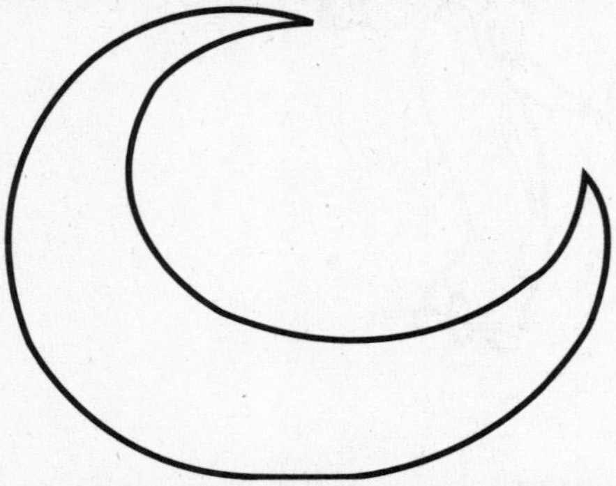

Moons in Space

Planets in Space

Constellations

Rotating Between Day and Night

Revolving Around the Sun

Earth

Seasons

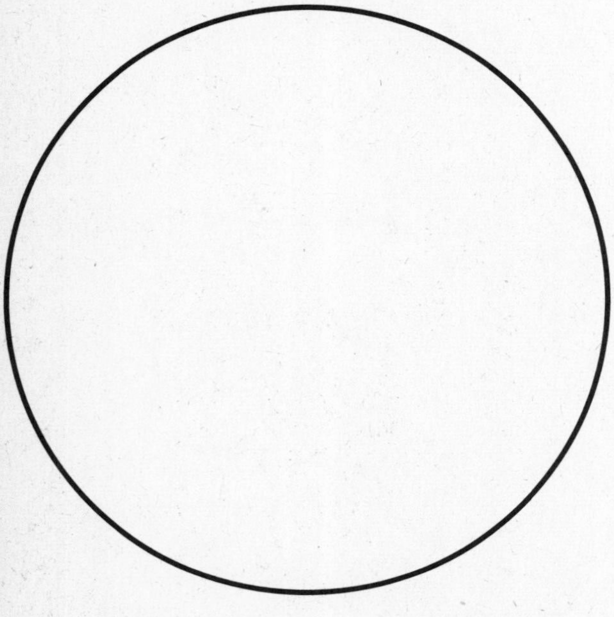

Use with p. 43
Revolving Around the Sun

The Moon's Surface

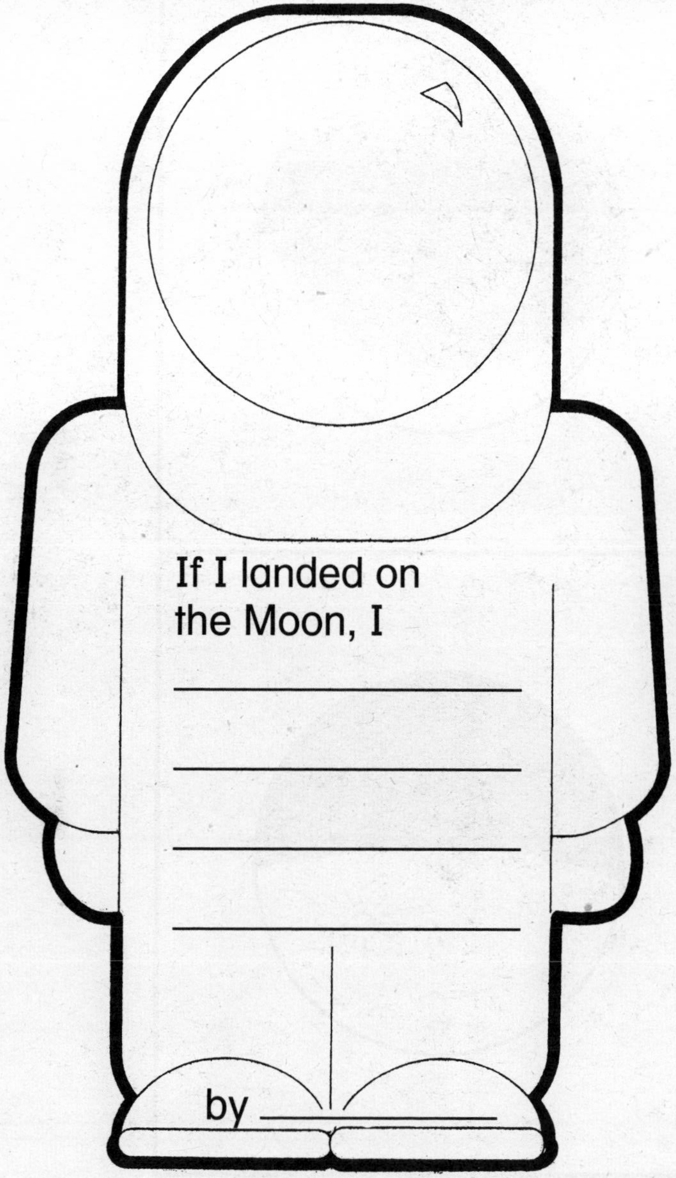
If I landed on the Moon, I
by

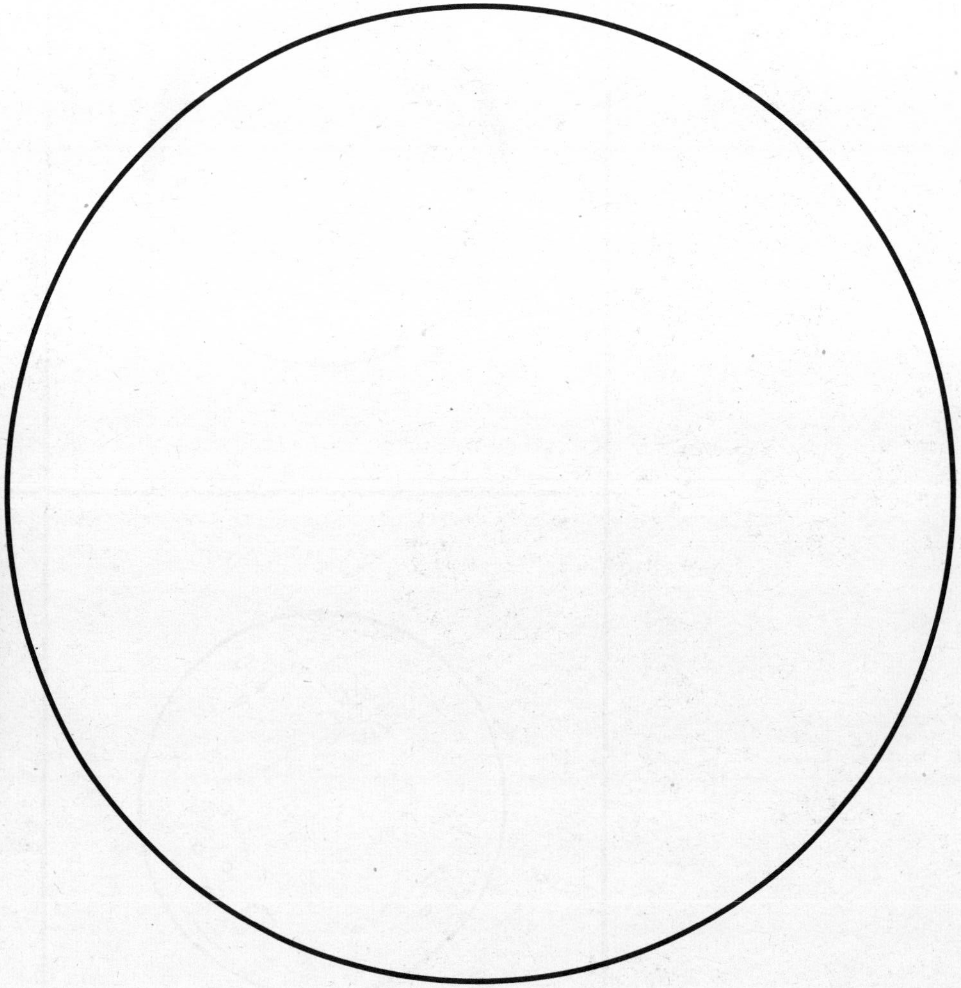

Phases of the Moon

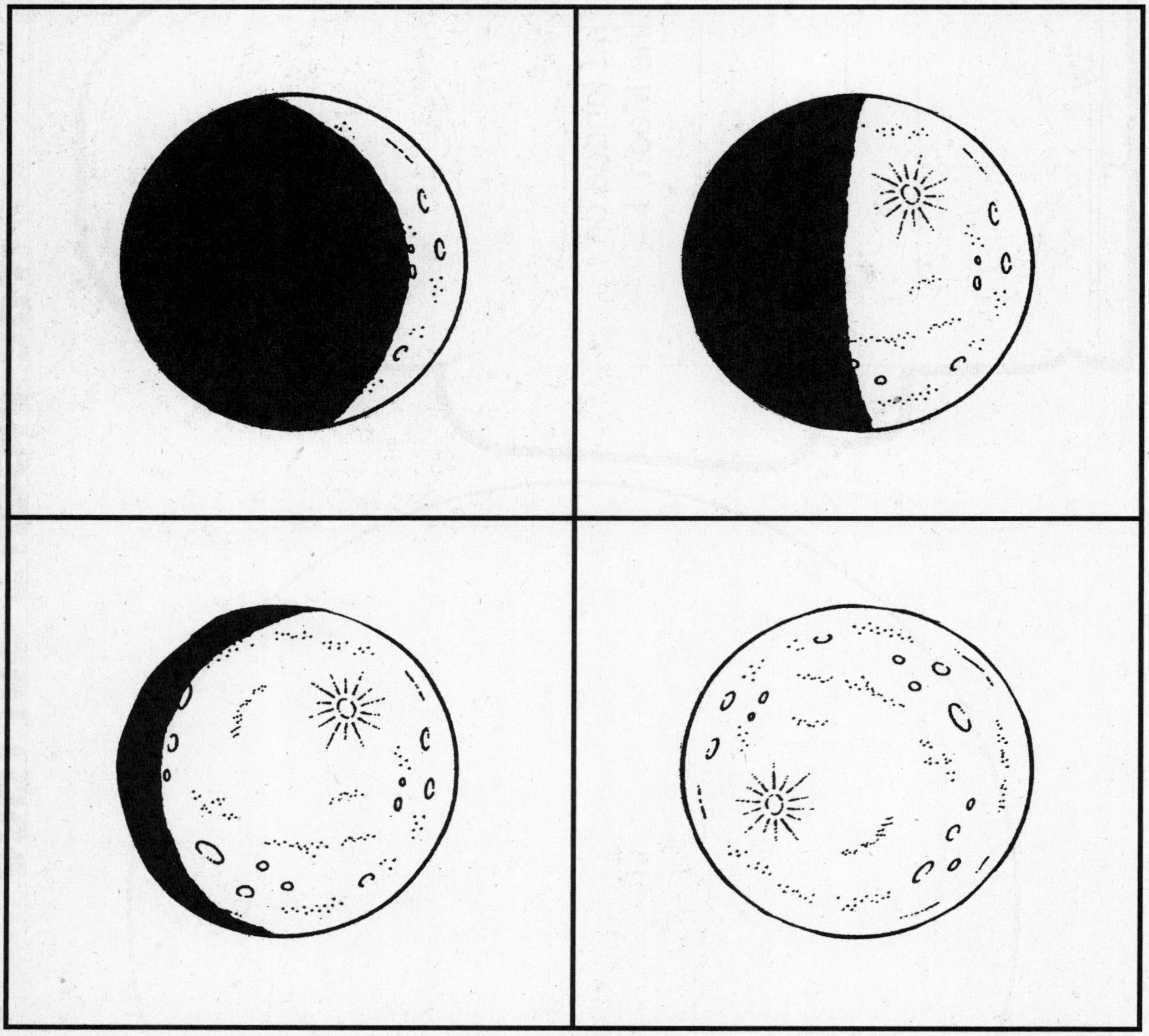

Use with p. 44
Phases of the Moon Quilt

Moon Exploration

1962

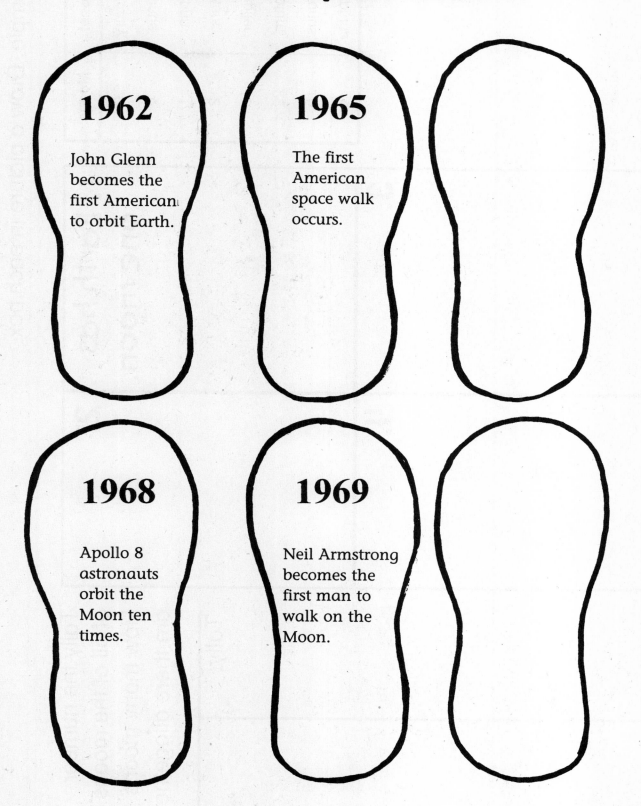

John Glenn becomes the first American to orbit Earth.

1965

The first American space walk occurs.

1968

Apollo 8 astronauts orbit the Moon ten times.

1969

Neil Armstrong becomes the first man to walk on the Moon.

Directions: Read the chart below. Write 3 sentences in the boxes. Box 1 is a sample. Draw a picture in each box.

Planet	Moons
Mercury	0
Venus	0
Earth	1
Mars	2
Jupiter	16
Saturn	22
Uranus	15
Neptune	8

1. Earth has one moon.	2.
3.	4.

Tally the number of all of the moons. How many moons are there altogether?

Tally:

Many Moons

Planets Go 'Round the Sun Song

(Tune: The Farmer in the Dell)

Refrain

The planets go 'round the Sun,
The planets go 'round the Sun.
Eight planets orbit the Sun,
The planets go 'round the Sun.

Verses

Venus has thick clouds,
Which hide the land below.
Shining brightly in the night,
I wish that I could go.

(Refrain)

Mars is next to Earth,
It's covered with red dust.
Spinning 'round and 'round in space,
We'll go there if we must.

(Refrain)

Saturn has many rings,
And 22 moons so bright.
Icy rocks in circles swirl,
They make a pretty sight.

(Refrain)

Neptune is in the sky,
Rotating 'round and 'round.
Covered with so many clouds.
More facts should be found.

(Refrain)

Mercury is hot,
And Mercury is small.
Mercury has no signs of life.
It's just a rocky ball.

(Refrain)

We live on Earth, our home,
With oceans and with trees.
It has clean water and air,
So no pollution, please.

(Refrain)

Great Jupiter is large,
And has a big red spot.
We know that it has 16 moons.
It's been studied quite a lot.

(Refrain)

Uranus is far away,
With 15 moons in view.
The closest planet to the Sun,
Is it green or blue?

(Refrain)

Planet Models

Follow the directions below in order to make models of all of the planets and the Sun.

Sun

Paint a paper plate yellow. When it is dry, make a pattern on it using a round bath sponge and orange paint. Curl eight orange paper strips by wrapping them around a pencil. Then glue them around the edge of the plate from the back.

Planets

Before making each model, read the directions for it below. Choose a lid that is the correct size and the right colored paper for each planet. Trace the lid on the colored paper and cut out the circle. Then decorate it by following the rest of the instructions.

Mercury (lid size 2) Use a toothbrush to brush black paint onto a red circle.

Venus (lid size 3) Use a sponge on a stick to add white paint to a yellow circle.

Neptune (lid size 3) Use a round bath sponge to stamp blue paint onto a blue circle.

Uranus (lid size 3) Brush green paint over bubble wrap and place a green circle over it and press down. Lift it off and let it dry. Then draw a silver ring across it.

Mars (lid size 4) Use a roller sponge to add red and orange paint to a red circle.

Saturn (lid size 4) Water down some paints and paint colorful stripes on a white circle. Draw silver rings across it.

Earth (lid size 4) Place small pieces of green tissue paper on a blue circle. Brush a glue and water mixture over the tissue paper and let it dry.

Jupiter (use a coffee filter) Put several colors of food coloring dye in separate cups of water. Use eye droppers to get small amount of the colored water, and then drop colors onto a white coffee filter.

The Solar System

Read this phrase to help you memorize the planets' order.

<u>M</u>y <u>V</u>ery <u>E</u>ager <u>M</u>om <u>J</u>ust <u>S</u>erved <u>U</u>s <u>N</u>achos

Write a new phrase using the first letters of the planets' names.

M_____ V_____ E_____ M_____ J_____ S_____ U_____ N_____.

____

____

Circle Patterns for Planets' Order

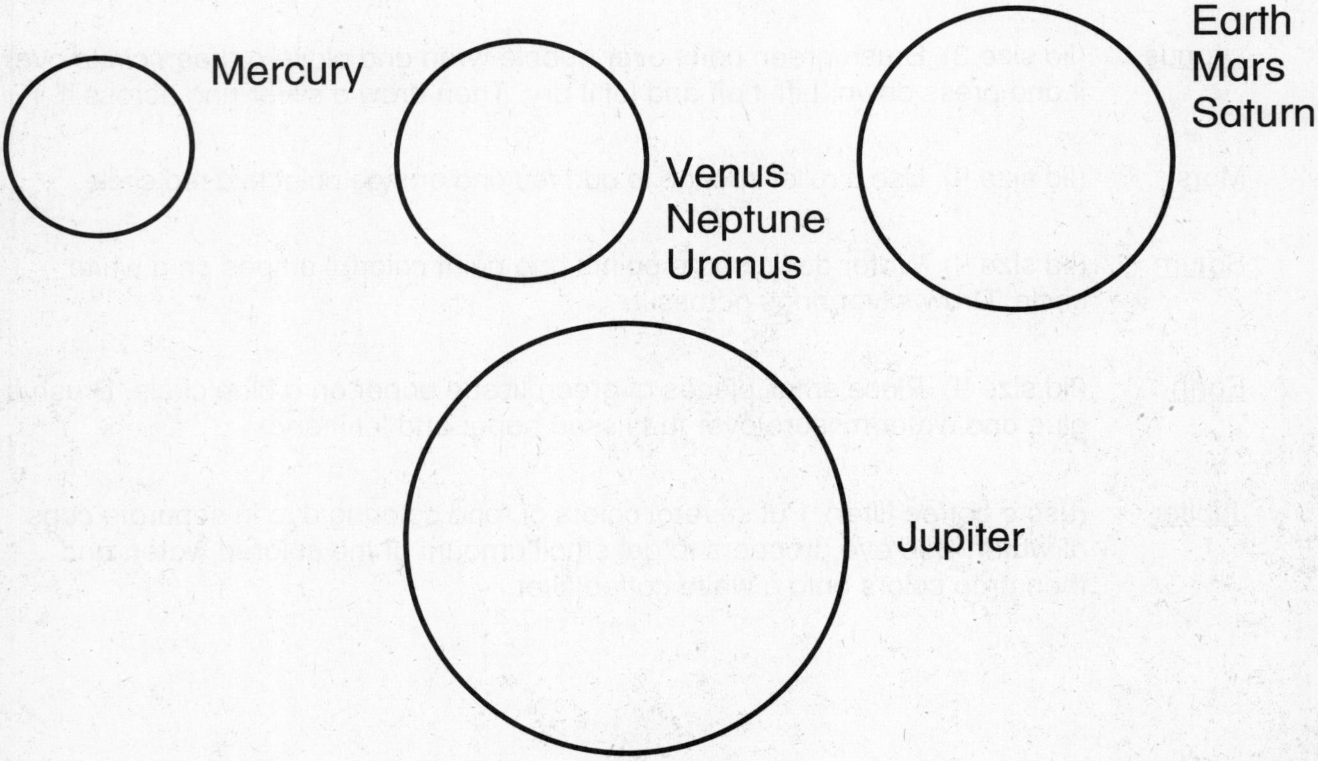

A ball of fire am I,
above you in the sky.
I give light and heat
and help grow things to
eat.

Eight of us
circle the Sun.
There's no life on us,
except for one.

I change my face
every night.
I'm full, half or crescent
shining like a star.

Starry Nights

Use with p. 48
Starry Nights

Constellations Galore

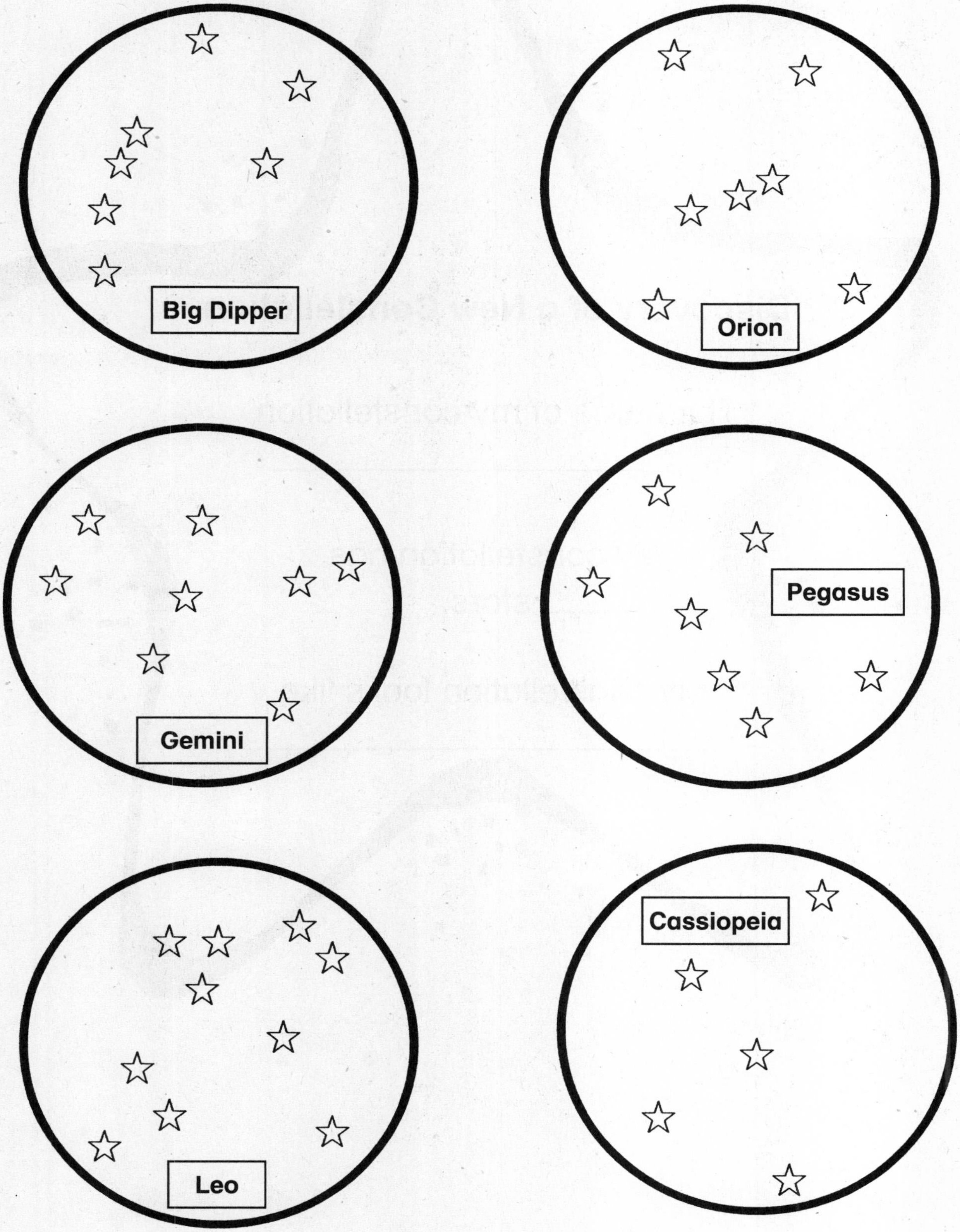

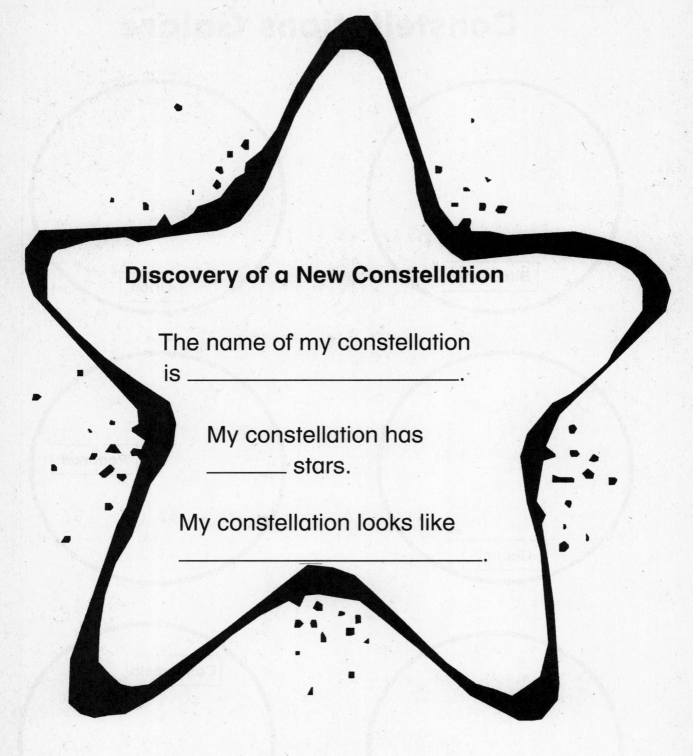

Discovery of a New Constellation

The name of my constellation is ______________________.

My constellation has _______ stars.

My constellation looks like ___________________________.

Use with p. 49
Creating a Constellations Quilt

Let's Study Matter

Changes in Matter

Let's Study Movement

Forces At Work

Use with p. 50

The color of the item is ______________________________.

The size of the item is ______________________________.

The item looks like ______________________________.

The item smells like ______________________________.

The item tastes like ______________________________.

The item feels like ______________________________.

Use with p. 50

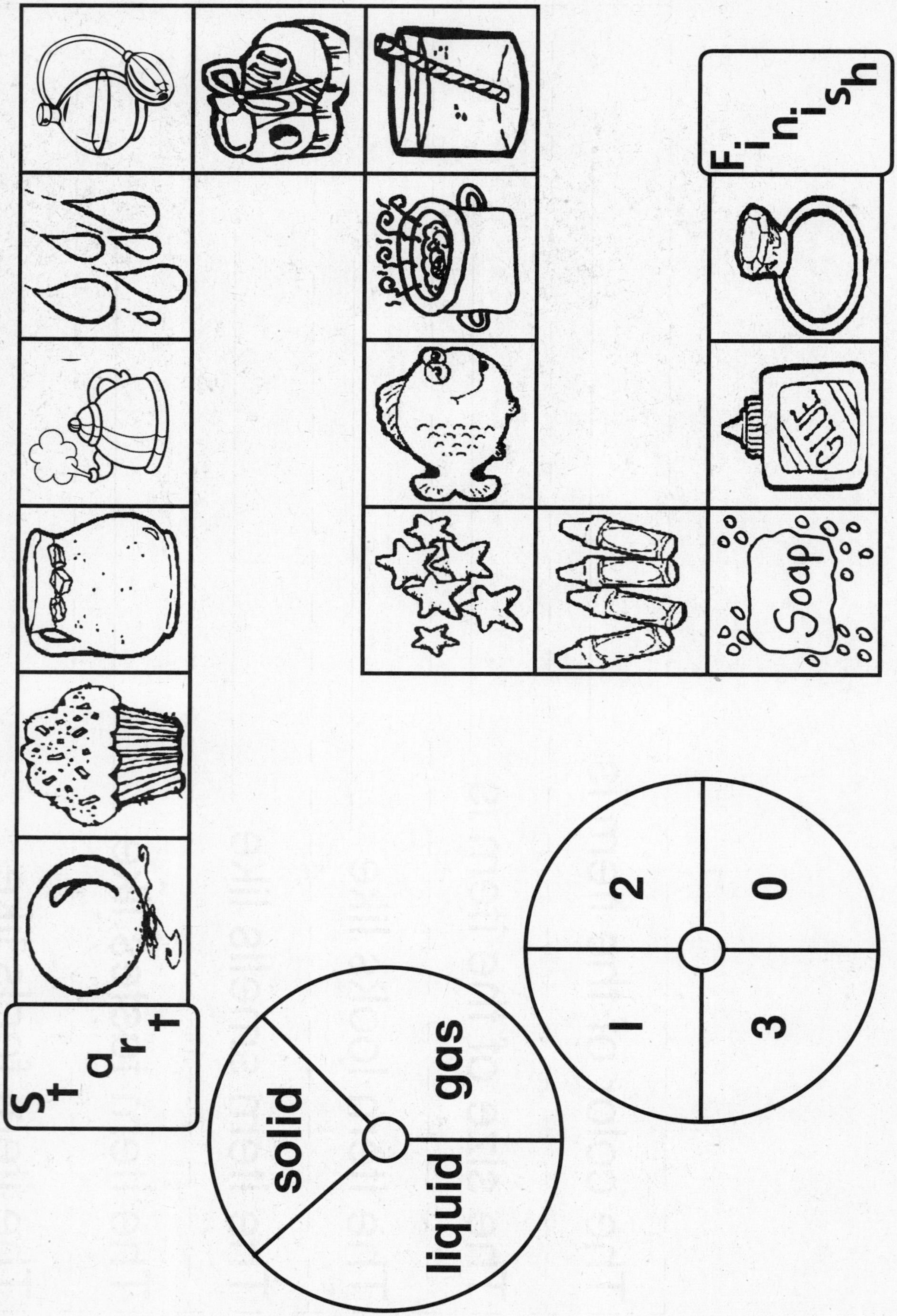

Use with **p. 50**

Earth is
matter.
So is the
sea.
The sky is
matter,
Just like you
and me.

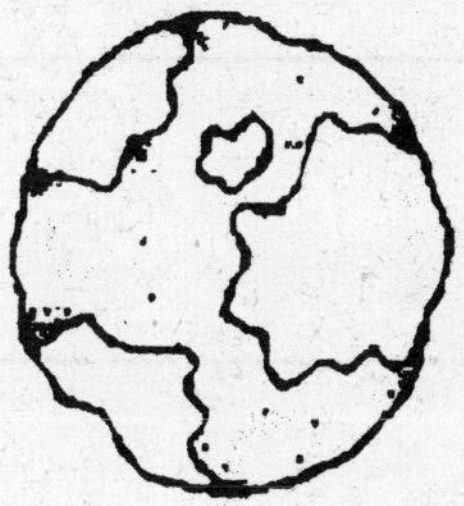

Matter Is All Around

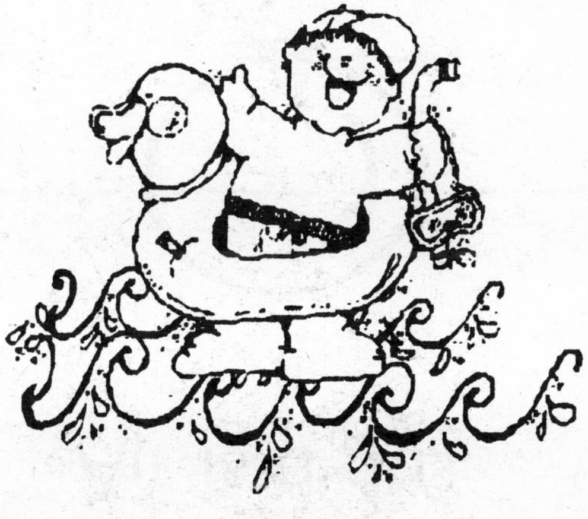

A chair is a solid.	Rain is a liquid.	Air is a gas.
		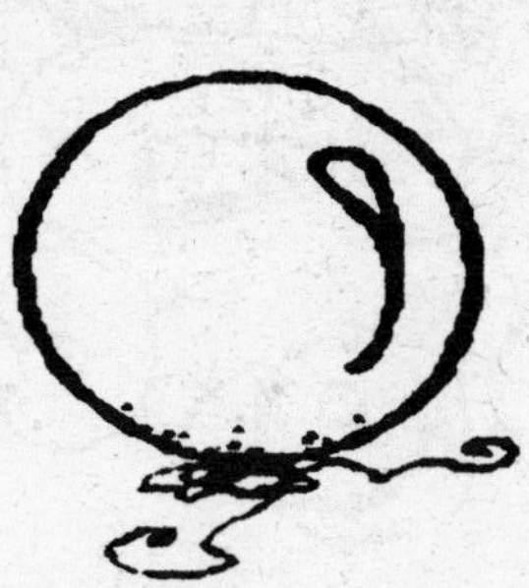
Other solids:	**Other liquids:**	**Other gases:**
____________	____________	____________
____________	____________	____________
____________	____________	____________
____________	____________	____________

Measuring Matter

	Connecting Cubes		Counting Bears		Counting Tiles		Linking Loops	
	Guess	Actual	Guess	Actual	Guess	Actual	Guess	Actual
Jar #1								
Jar #2								
Jar #3								

Bread With Butter

Toast With Melted Butter

Heat Changes Matter

Choose a type of food that changes when you add heat to it. In the space below, draw and color a picture of that food before heat is added.

In the space below, draw and color a picture of that food after heat is added.

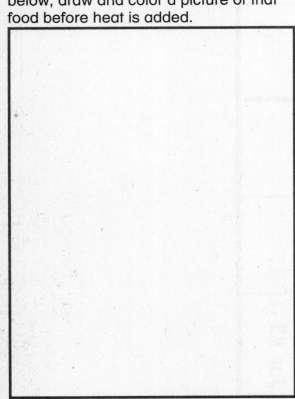

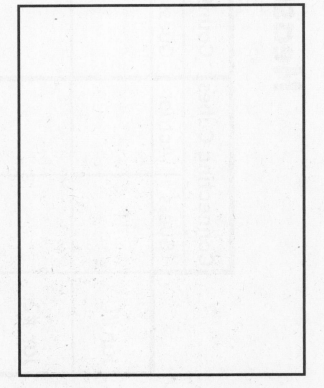

Ice Cream Recipe

Mix $1\frac{1}{2}$ cups of sugar with 1 small can of evaporated milk, 2 teaspoons of vanilla, 1 package of vanilla instant pudding, and 6 cups of milk.

Put $\frac{1}{3}$ cup of mixture into a small plastic bag. Seal the bag tightly. Fill a large plastic bag with ice.
Add $\frac{1}{2}$ cup of salt to the ice.

Cooling Matter

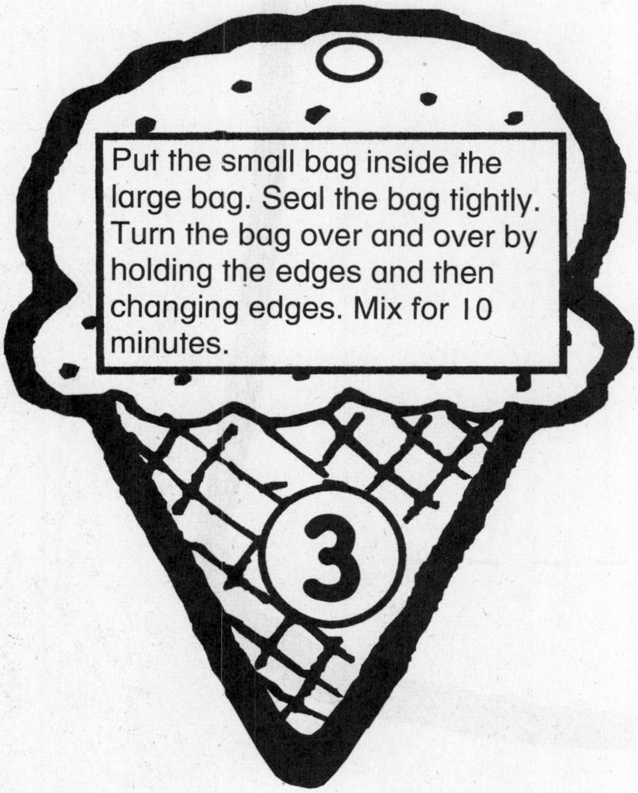

Put the small bag inside the large bag. Seal the bag tightly. Turn the bag over and over by holding the edges and then changing edges. Mix for 10 minutes.

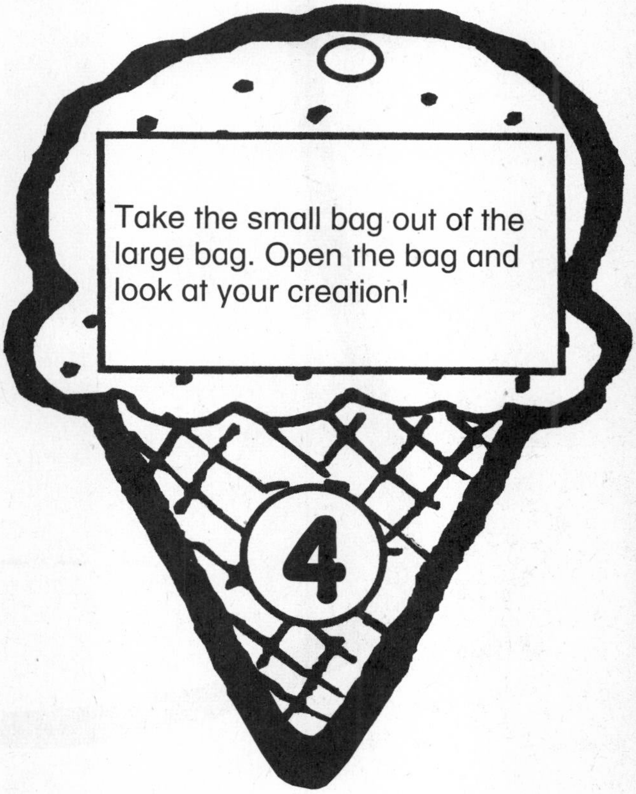

Take the small bag out of the large bag. Open the bag and look at your creation!

Changes in Matter

water

↓

steam

liquid
to
gas

ice cubes

↓

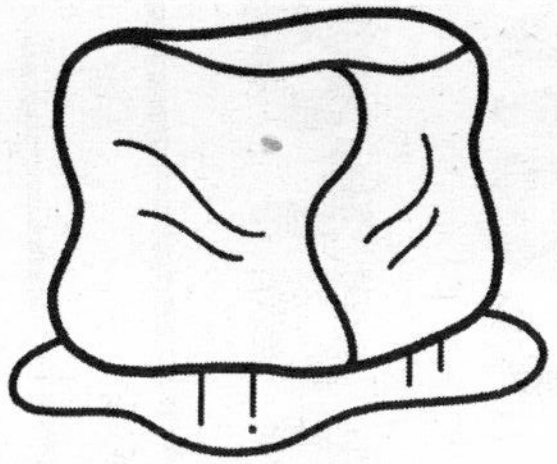

ice melting

solid
to
liquid

balloon

↓

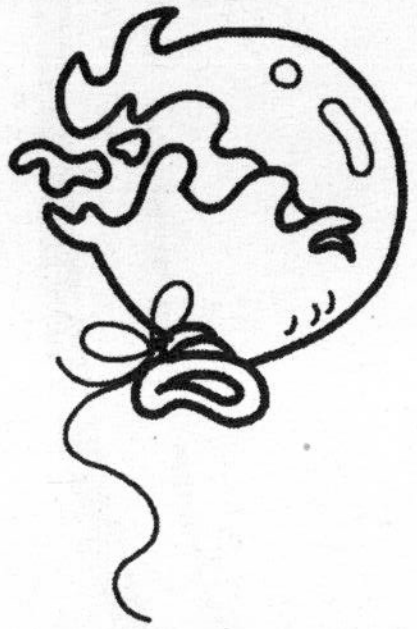

broken balloon

gas
to
gas

cake mix

↓

baked cake

liquid
to
solid

I can see
separate parts.

I cannot see
separate parts.

Use with p. 53
Bowls of Matter

Trail Mix	Salad
Pudding	Cake Mix
Cookies	Cereal
Stew	Ice Cream
Trail Mix	Salad
Pudding	Cake Mix
Cookies	Cereal
Stew	Ice Cream

The 3 Bears' Neighborhood

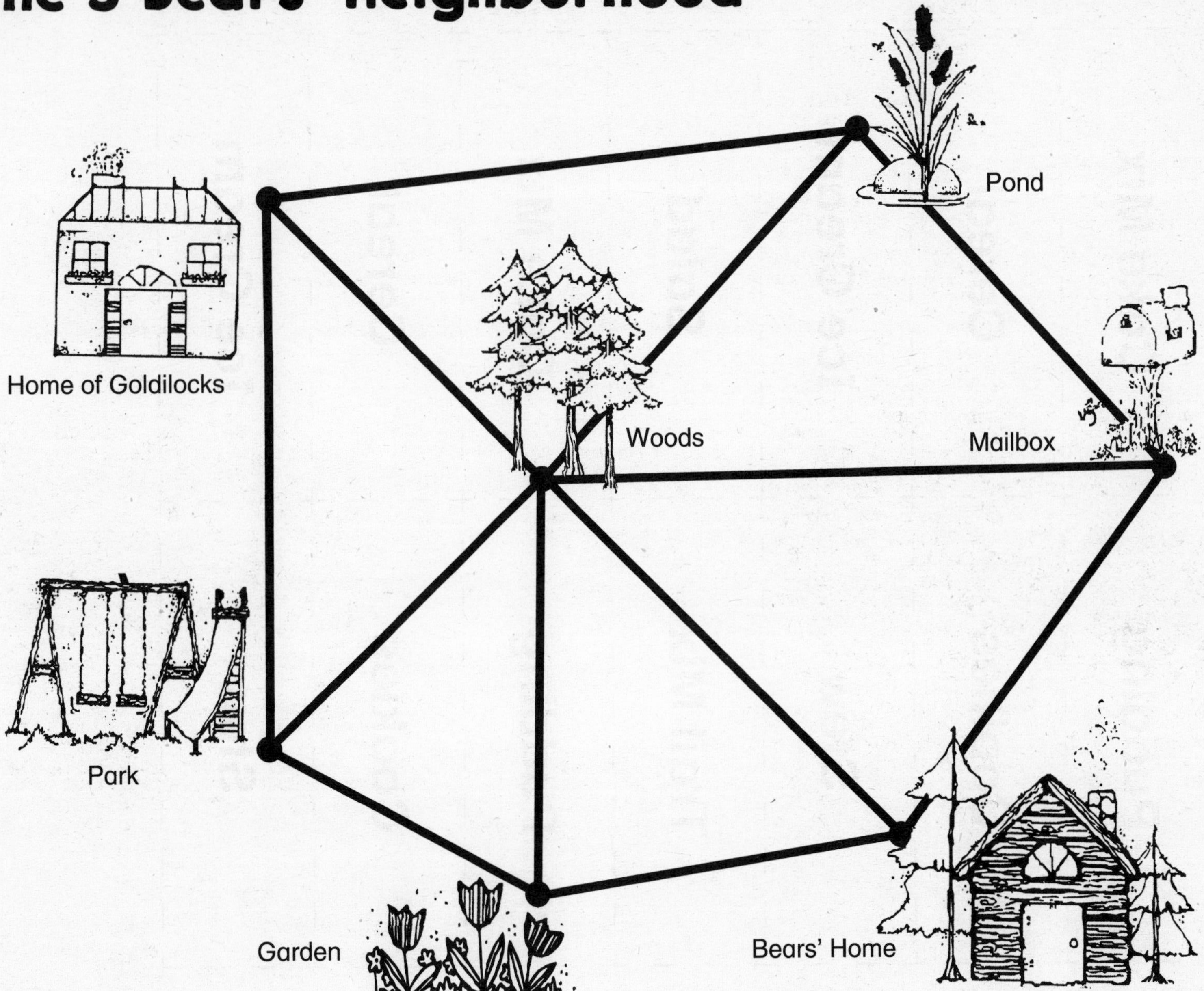

Adventures in The 3 Bears' Neighborhood

Use a ruler to figure out the distance between objects, measuring from dot to dot.

Trip	Distance on Map
1. Mama Bear walked from the Bears' Home to the Garden to pick some flowers.	2.5 inches
2. Papa Bear walked from the Bears' Home to Goldilocks' house to invite her to Baby Bear's birthday party. He then walked home again.	11 inches
3. Baby Bear walked from the Bears' Home to the Woods and then to the Mailbox. After picking up the mail, he walked back home.	10 inches
4. Goldilocks left her house and walked to the Pond to count the frogs. She then went on to play at the Park with her friends.	9.5 inches
5. Baby Bear left his house and walked to the Woods and then to the Park.	5.5 inches
6. What is the shortest path to the Home of Goldilocks from the Bears' Home?	through the Woods

I rolled my car across ______________.

My car went ______________.

I rolled my car across ______________.

My car went ______________.

I rolled my car across ______________.

My car went ______________.

I rolled my car across ______________.

My car went ______________.

Use with p. 55

Directions: I. Write the name of one of these objects in each square:
wagon, sled, trailer, stroller, swing, cart, mower, water-skier.

2. Flip a penny. Heads: Cover an item that is pushed.
Tails: Cover an item that is pulled.

Push-or-Pull Blackout

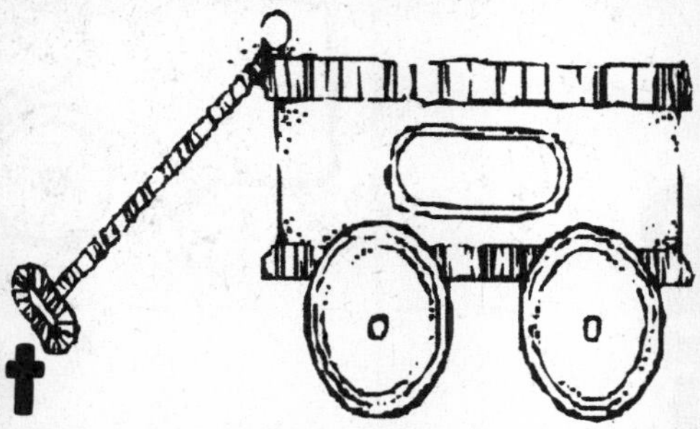

	Heads (push) Tails (pull) **Free**	

Machine Match-Up

 wheel and axle	 wheel and axle	 wheel and axle	 wheel and axle
 wedge	 wedge	 wedge	 wedge

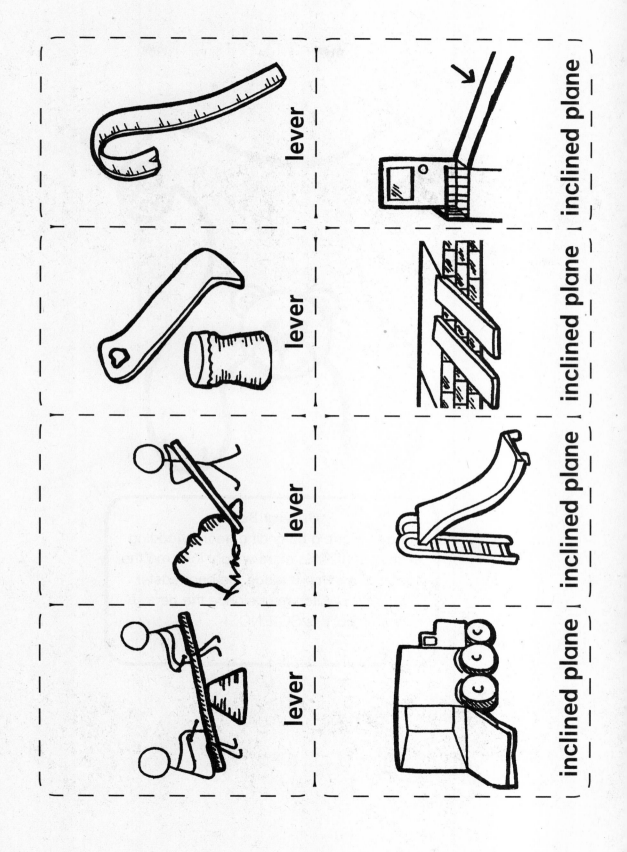

 lever	 lever	 lever	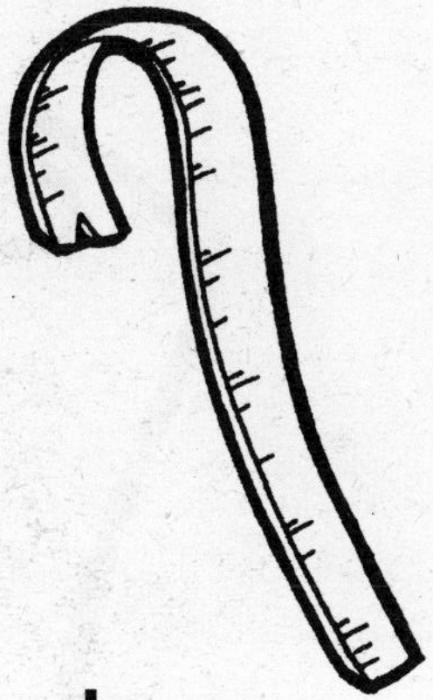 lever
 inclined plane	 inclined plane	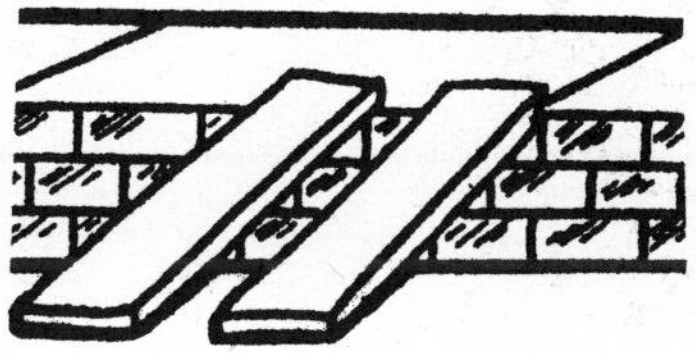 inclined plane	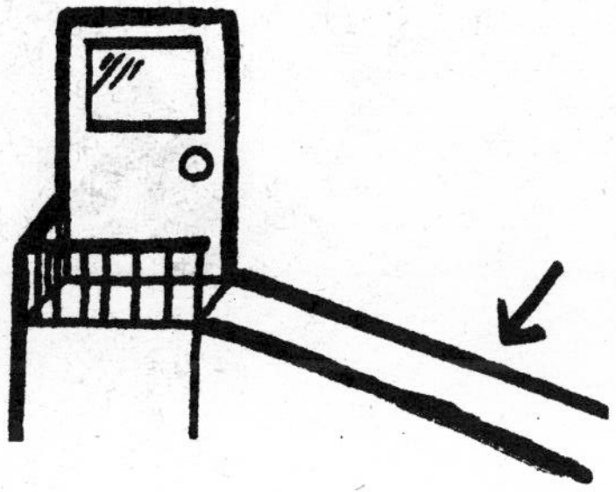 inclined plane

Gravity Bear

Today I made a balancing bear by adding extra weight. This extra weight lowered the center of gravity. It made the bear easier to balance. Watch me balance the bear. It is not magic, it is SCIENCE!

Feature Attractions

by ____________________

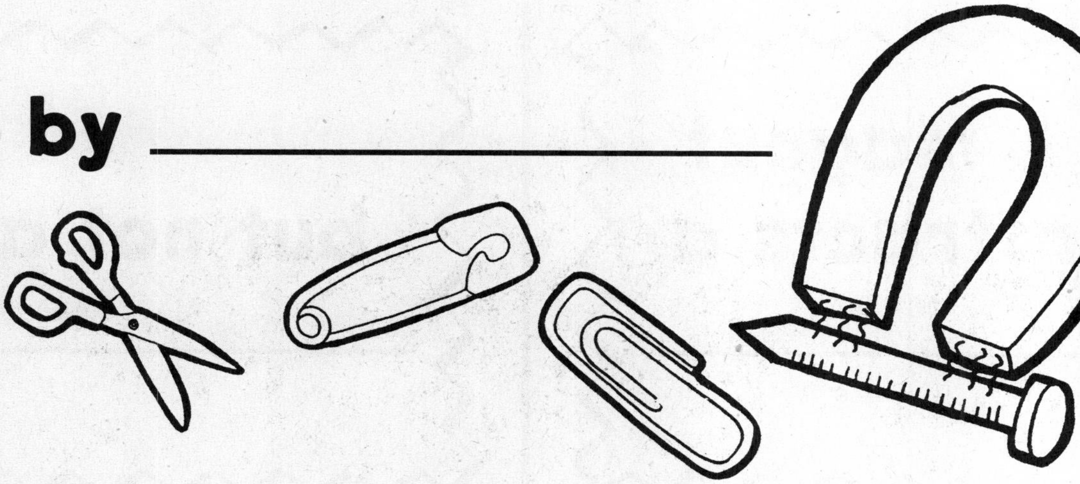

A magnet attracts a

____

but not a

____________________.

A magnet attracts a

____

but not a

____________________.

Soft Sounds

Loud Sounds

Directions: The following items are in the jars: toothpicks, rice, dice, pennies, paper clips, pebbles, pom poms, and marbles. Match jars labeled #1–8 with jars labeled A–H. After the numbers, write your guess of the matching letters.

Heat

Sound

Light

Electricity

The Sun Warms Everything

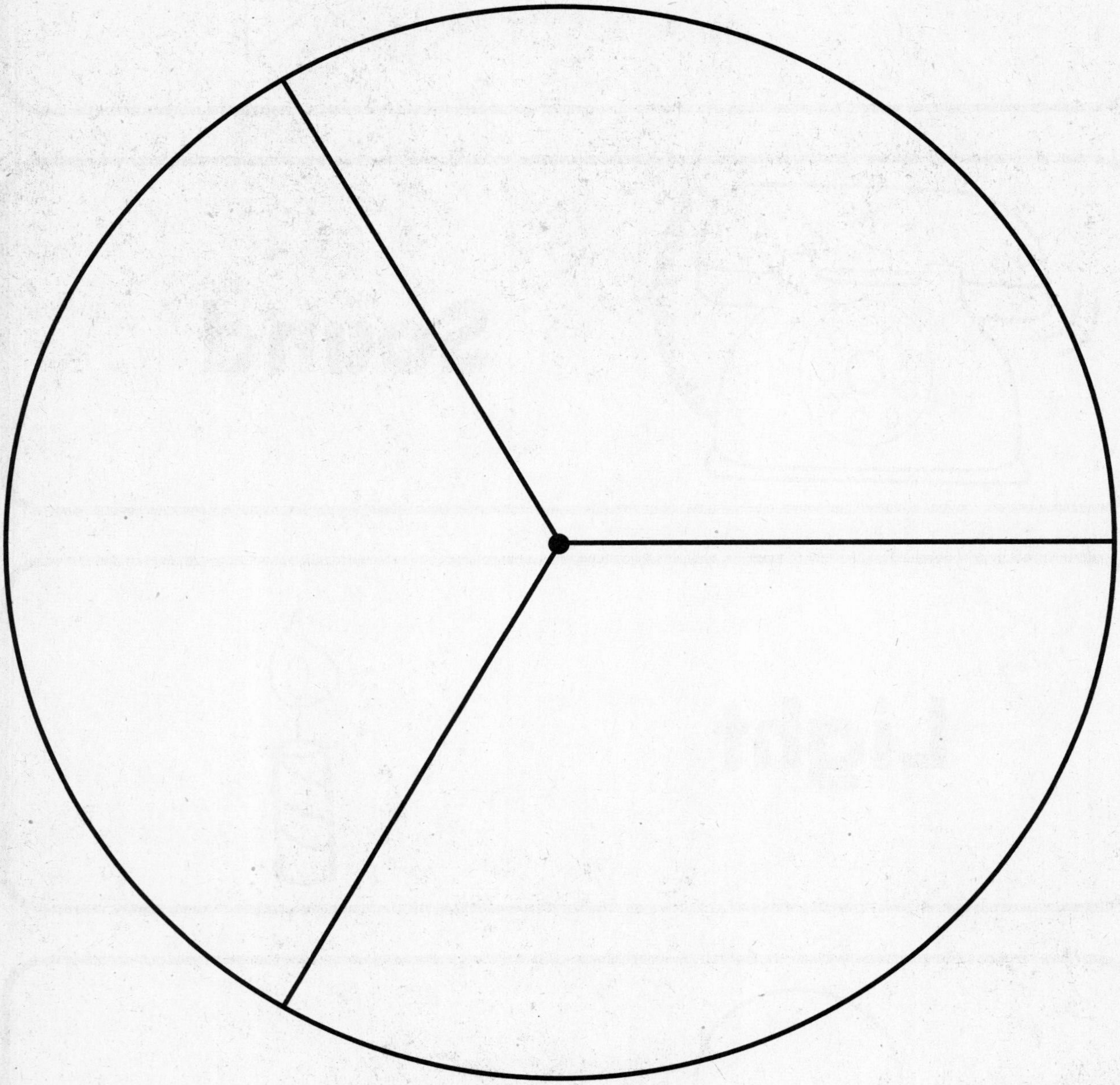

Changing Temperatures

Riddle 1

The coldest temperature is next to the hottest temperature.
The temperature inside a freezer is first.
The temperature inside a house is not last.
The temperature on a snowy day is not next to the hottest.

(0, 90, 68, 32 degrees Fahrenheit)

Riddle 2

The freezing temperature is not in the middle.
The hottest temperature is on one of the ends.
The warm temperature is to the left of the hottest.
The temperature of snow is next to the coldest.

(0, 32, 68, 90 degrees Fahrenheit)

Riddle 3

The temperature that makes you sweat is not in the middle.
The temperature that makes ice cubes is either third or fourth.
The temperature that makes you feel warm in is either first or second.
The temperature that chills your ice cream is last.

(90, 68, 32, 0 degrees Fahrenheit)

Riddle 4

The one digit number is to the left of the highest number.
The number that is 30 plus 2 is to the right of 90.
The number that is 10 less than 100 is not first.
The number that is closest to 70 is to the left of the smallest.

(68, 0, 90, 32 degrees Fahrenheit)

Containing Body Heat

It's cold outside as I get ready to play,
But I'll stay warm in the snow today.

I have a heavy ______________ to wear,
So if it's cold I don't really care.

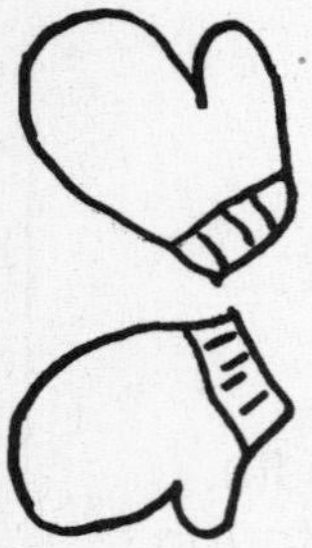

I put some ______________ on my hands,
They are soft and warm with stretchy bands.

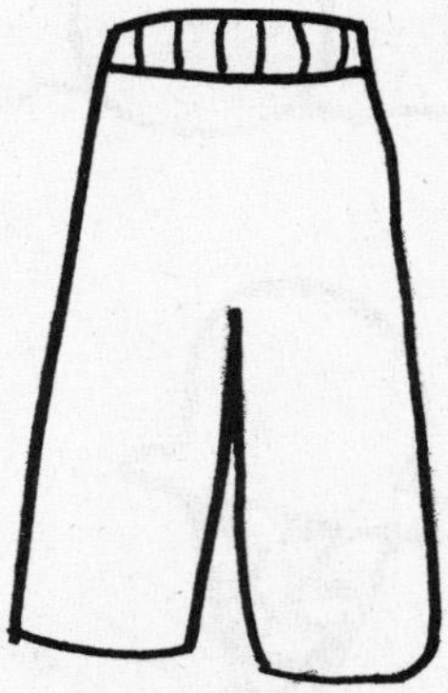

I'll put some ______________ on my feet,
With fur on them, they sure look neat.

A colorful ______________ goes under my chin,
And you can hardly see my big grin.

I'll put on ______________ that are so long,
So I can run, I really am very strong.

And on my head I'll wear a ______________,
Now what do you think of all of that?

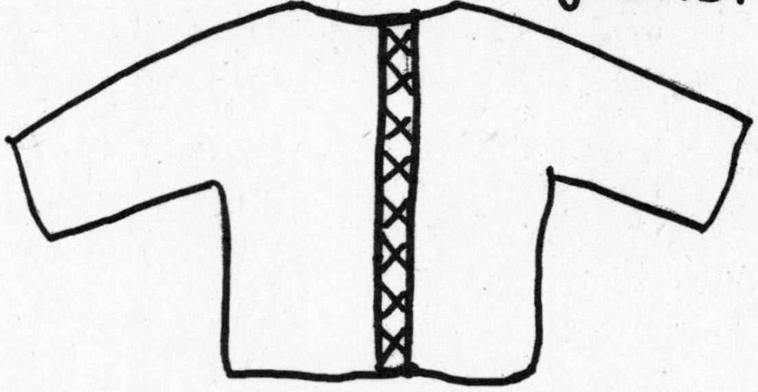

Manmade Sounds

Nature Sounds

Environmental Sounds Card Game

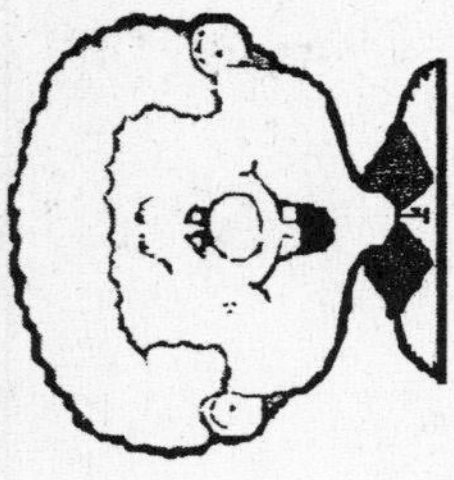

hoot,
hoot

ha, ha, ha

pop,
pop

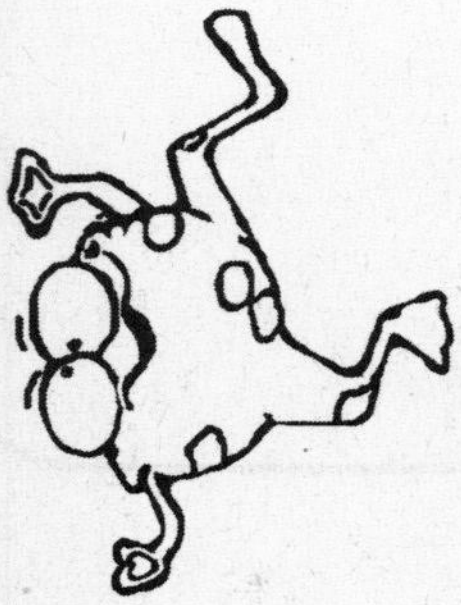

ring,
ring

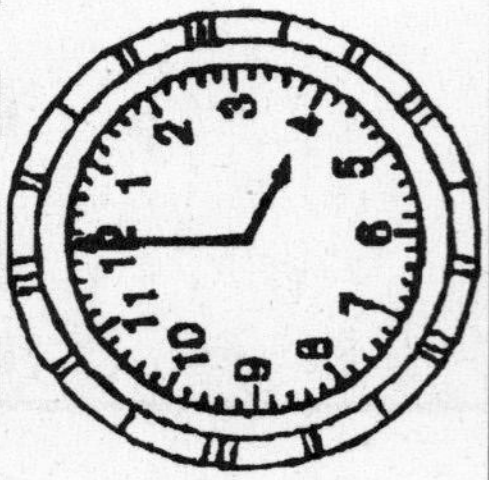

ribbet,
ribbet

Hi, I'm
Tom.

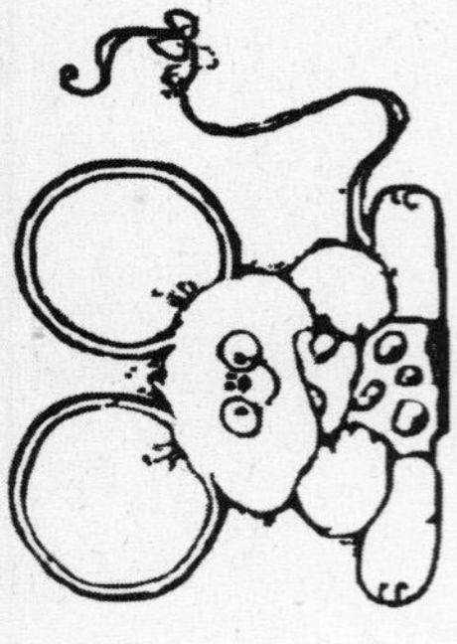

tick, tock
tick, tock

moo,
moo

buzz,
buzz

meow,
meow

drip, drop
drip, drop

squeak,
squeak

Use with p. 60
Environmental Sounds Card Game

Environmental Sounds Card Game

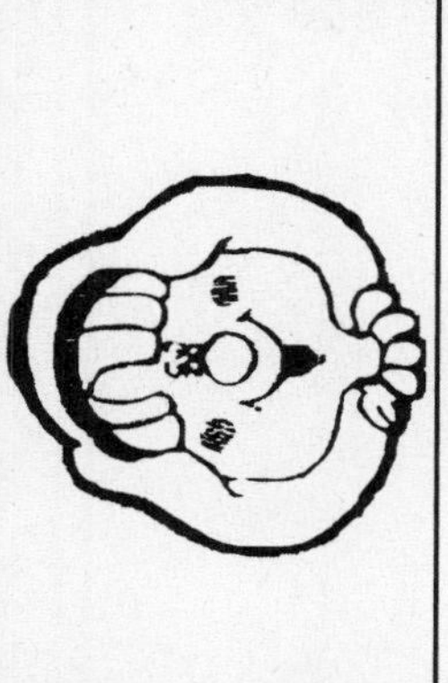

baa, baa	Hi, I'm Annie.	vroom, vroom	quack quack

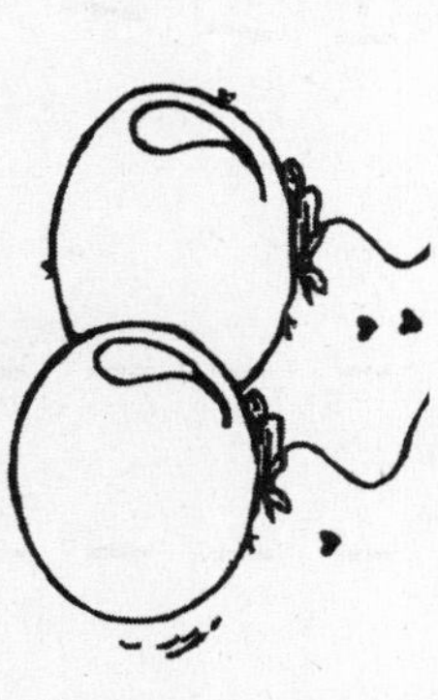

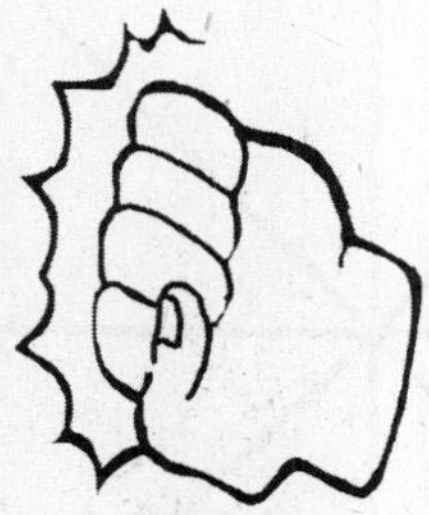
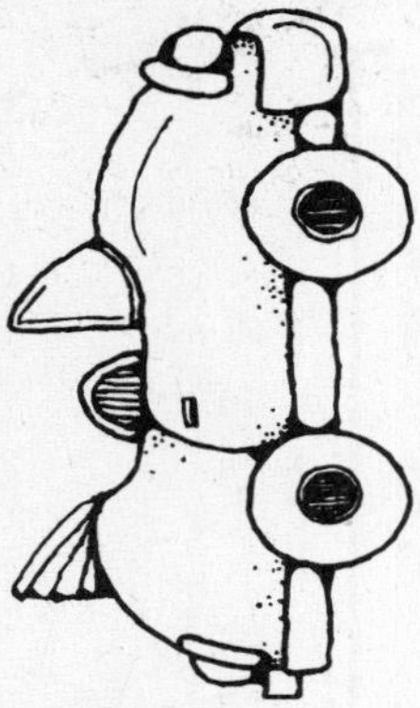

boo, boo	vroom, vroom	tweet, tweet	gobble, gobble

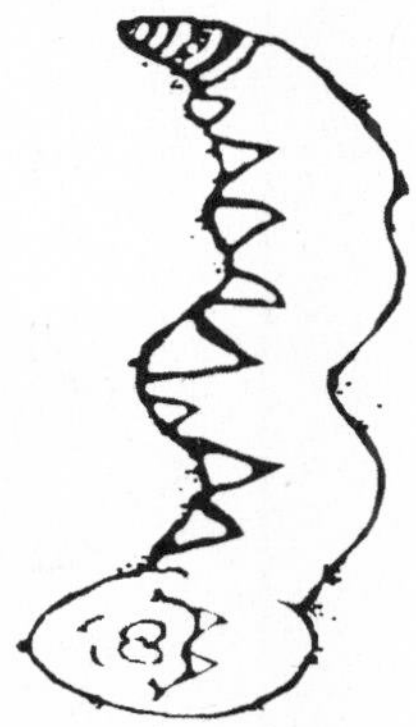

zoom, zoom	hiss, hiss	Happy birthday!	oink, oink

Loud and Soft Sounds

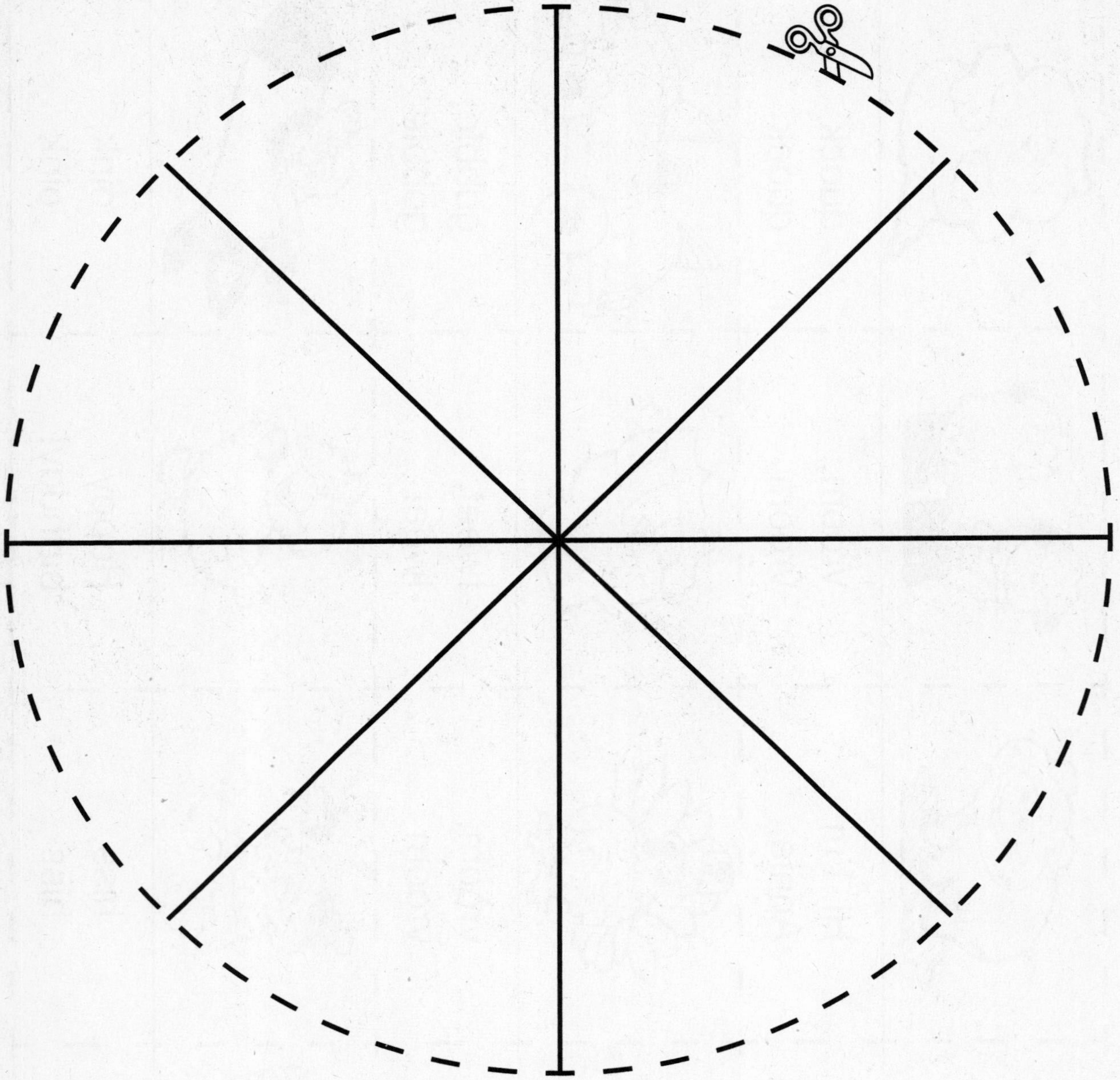

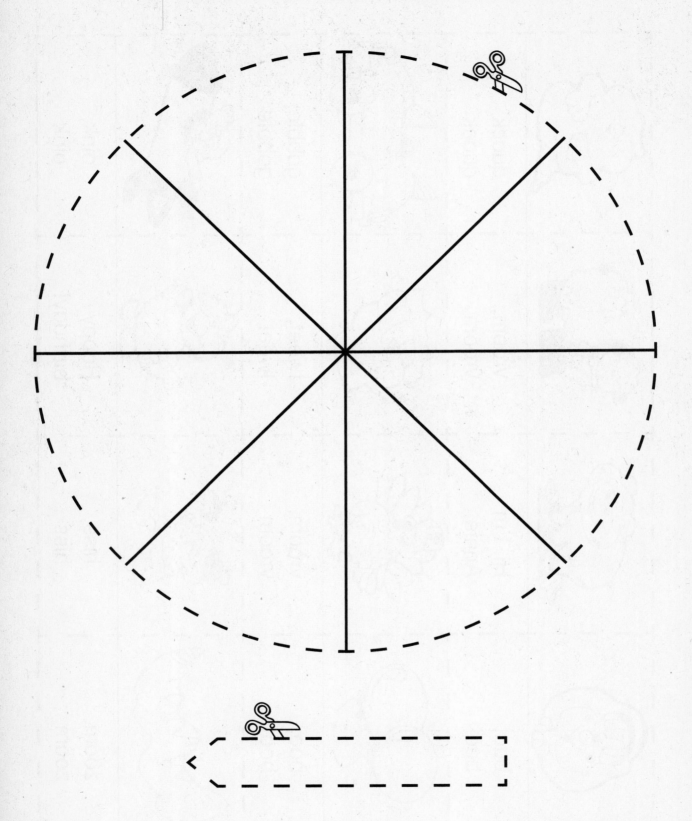

Loud and Soft Sounds

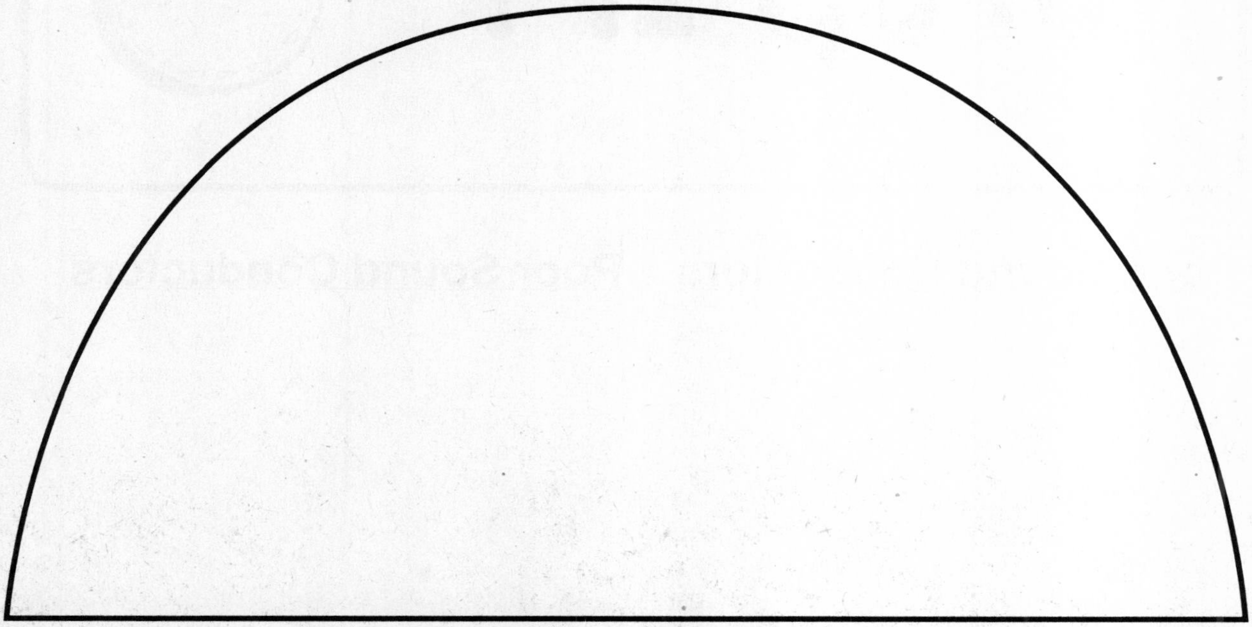

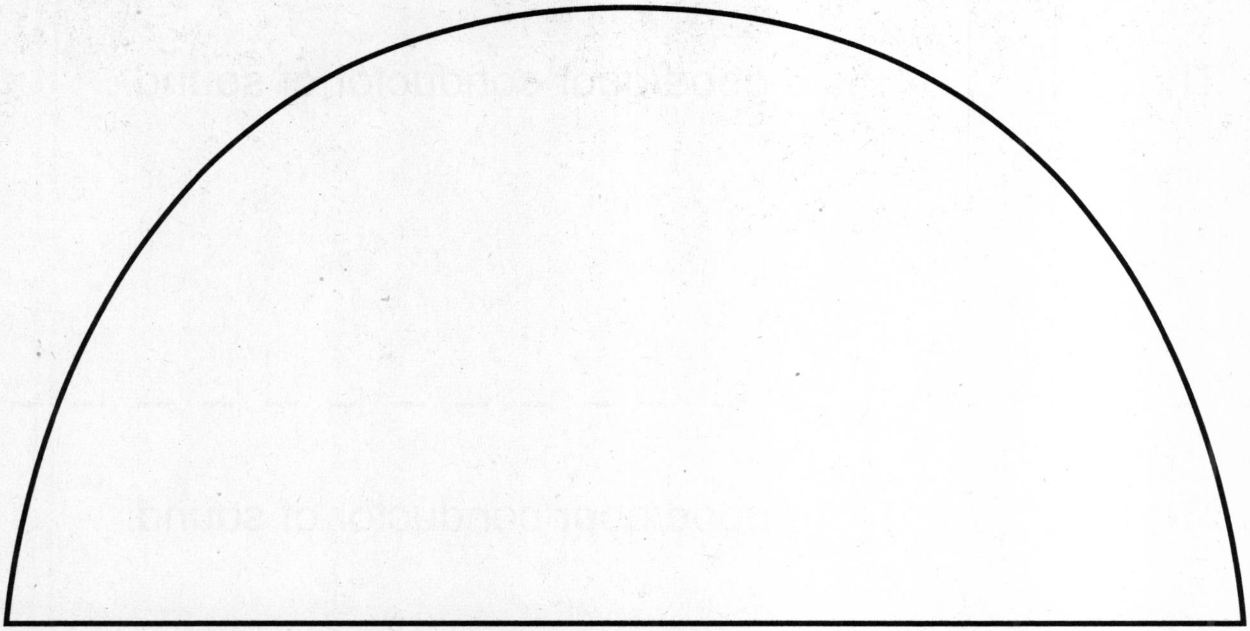

Drop It !

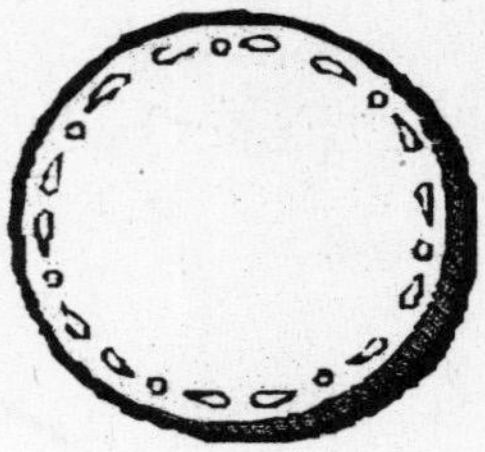

<u>Good Sound Conductors</u> **<u>Poor Sound Conductors</u>**

The __________ is a good/poor conductor of sound.

The __________ is a good/poor conductor of sound.

The __________ is a good/poor conductor of sound.

The __________ is a good/poor conductor of sound.

The __________ is a good/poor conductor of sound.

The __________ is a good/poor conductor of sound.

Sources of Light

Sources of Light

Light Spectrum

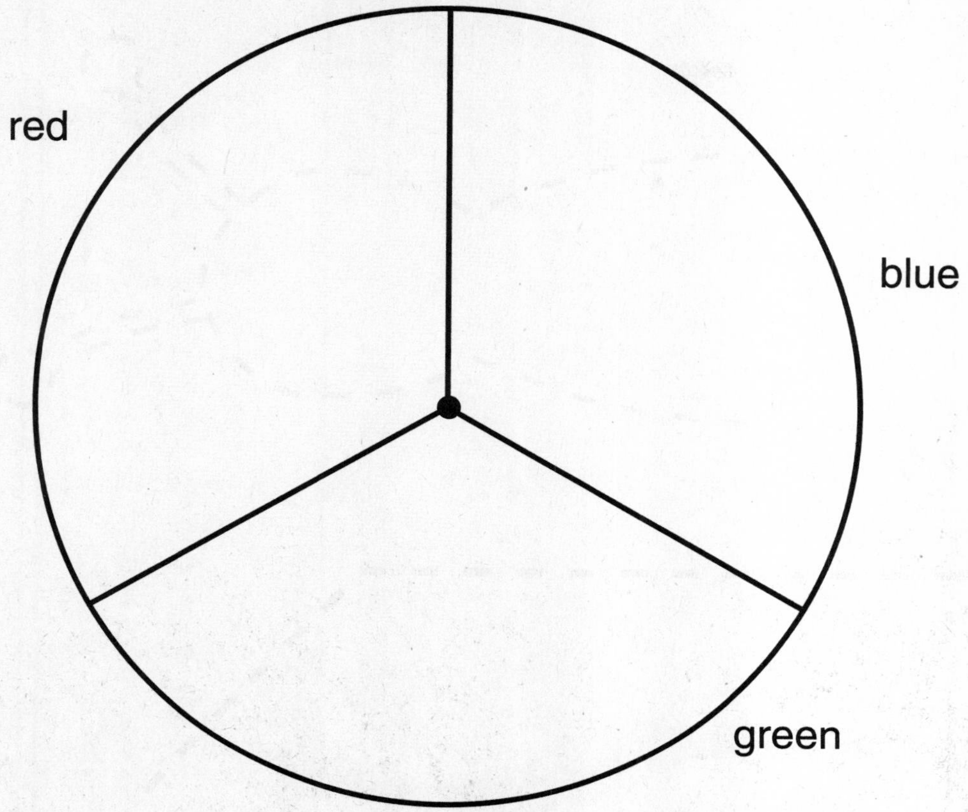

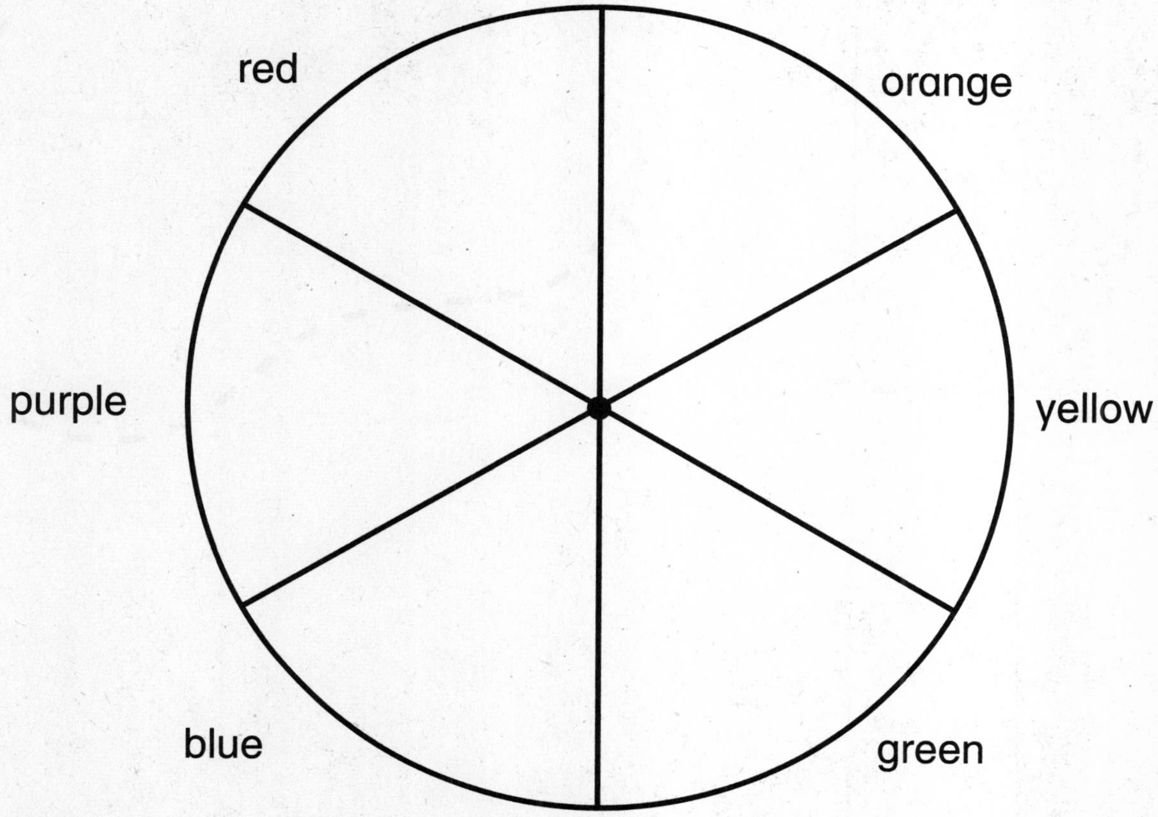

Static Electricity

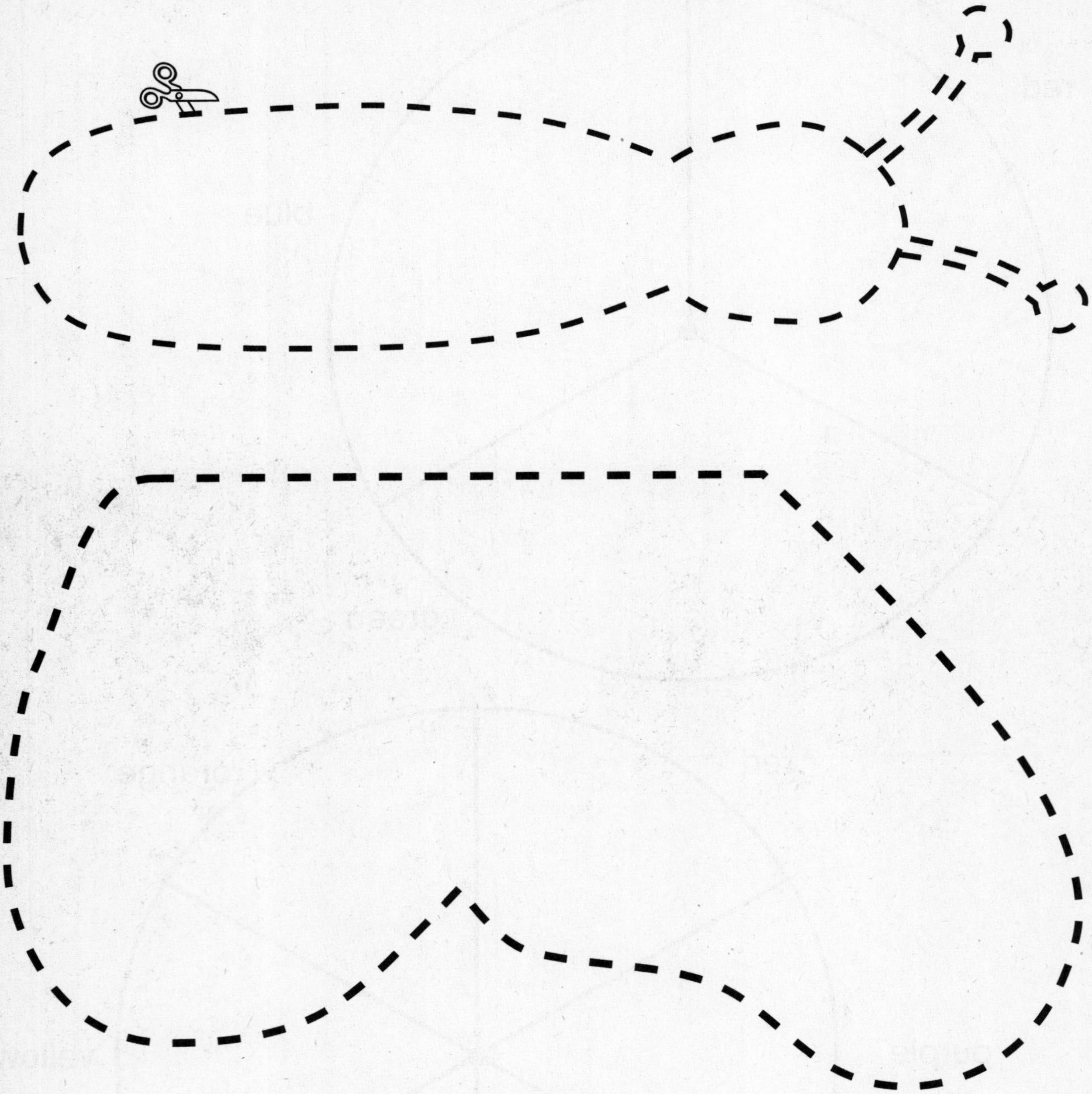

Inventors and Electricity

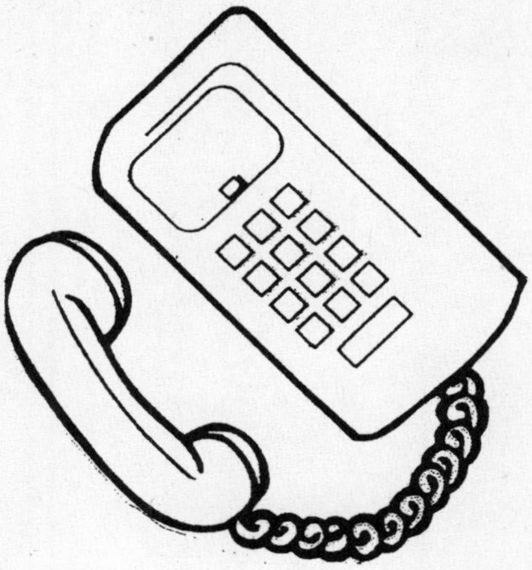

Benjamin Franklin discovered that lightning is electricity.

Alessandro Volta made the first battery that produced a steady flow of electric current.

Alexander Graham Bell invented the first telephone.

Thomas Edison discovered ways to make a light bulb burn longer and brighter.

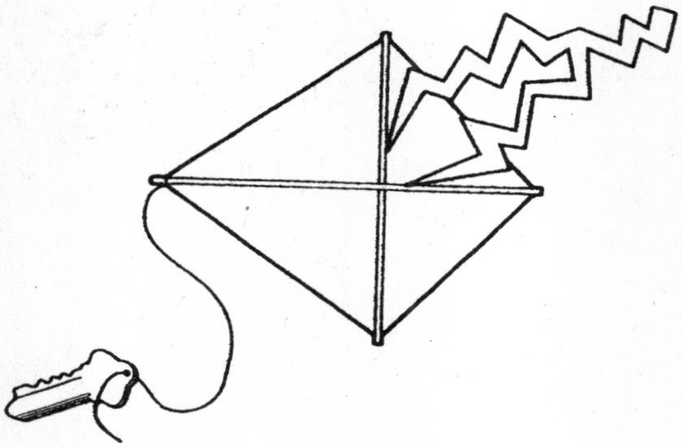